International Manual of Climate Change Control (IMCCC)

For all people who wish to take care of climate change

First Edition

Jahangir Asadi
Vancouver, BC CANADA

Copyright © 2022 by Top Ten Award International Network.

All rights reserved. No part of this publication may be reproduced, distributed or transmitted in any form or by any means, including photocopying, recording, or other electronic or mechanical methods, without the prior written permission of the publisher, except in the case of brief quotations embodied in critical reviews and certain other noncommercial uses permitted by copyright law. For permission requests, write to the publisher, addressed "Attention: Permissions Coordinator," at the address below.

Published by: Top Ten Award International Network
Vancouver, BC **CANADA**
Email: Info@TopTenAward.net
www.TopTenAward.net

Ordering Information:
Quantity sales. Special discounts are available on quantity purchases by universities, schools, libraries, corporations, associations, and others. For details, contact the "Sales Department" at the above mentioned email address.

International Manual of Climate Change Control (IMCCC)/J.Asadi—First Edition

ISBN 978-1-990451-50-8   Paperback

ISBN 978-1-990451-51-5   Hardcover

Contents

About TTAIN 10

Environmental Sustain for Future Kids (ESFK) 12

Introduction 13

Step one: Food & Recycling 15

Step Two: Energy 35

Step Three: Fashion & Textile 53

Step Four: Health & Beauty 83

Step Five: Cleaning 107

Step Six: Stationery 131

Step Seven: Do It Yourself (DIY) 155

Step Eight: Agriculture & Gardening 179

Step Nine: Professional 203

Step Ten: Financial 223

Step Eleven: Tourism 243

Step Twelve: Knowledge Test & Receiving Certificate 254

Bibliography 256

Acknowledgments:

I wish to thank my great friend, Mr. Habib Jamshidian who is the manager at H&B Canada Immigration Services for all of his support and sponsorship.

Acknowledgments:

I wish to thank my committee members, who were more than generous with their expertise and precious time. I would like to acknowledge and thank the Top Ten Award International Network for allowing me to conduct my research and providing any assistance requested.

It should be noted that all the required permissions for using the logos and trade marks has been obtained to be published in this volume.

Note:
We've done our absolute best to provide the best information possible, but since we haven't tried every single one of these solutions in every possible situation, we can't vouch for them 100 percent.

RESPECT THE
ENVIRONMENT

AFFORDABLE AND CLEAN
ENERGY

GO GREEN FOR THE
FUTURE

PLANT
MORE TREES

About TTAIN

Top Ten Award International Network

Top Ten Award international Network (TTAIN) was established in 2012 to recognize outstanding individuals, groups, companies, organizations representing the best in the public works profession.
TTAIN publishing different books related to Climate Change Control, international Eco-labeling, Management Systems, Food Safety, Medical Equipment management system, plans to increase public knowledge in taking care of our planet.
Top Ten Award International Network provides A to Z book publishing services and distribution to over 39,000 booksellers worldwide, including Apple, Amazon, Barnes & Noble, Indigo, Google Play Books, and many more.
Our services including: editing, design, distribution, marketing
TTAIN Book publishing are in the following categories:
Student
Standard
Business
Professional
Honorary

We focus on quality, environmental & food safety management systems as well as environmental sustain for future kids.

TTAIN also provide complete Online Training services for Climate Change Control courses, QMS, EMS, FSMS, HACCP and Eco-labelling based on international standards.

TTAIN has enough experiences to help create new ecolabeling programmes in different countries all over the world.
For more detail visit our website :
http://toptenaward.net   and/or   http://toptenaward.org
(Online Training courses) and/or
send your enquirer to the following email:
info@toptenaward.net

Global

Environmental Sustain for Future Kids established in Vancouver, BC Canada in 2020. (ESFK) is an international ecolabel focused on taking care of environment for future of kids.

ESFK defined as 'self-declared' environmental claims made by manufacturers and businesses based on ISO 14020 series of standards, the claimant can declare the environmental objectives and targets in relation to taking care of environment for future kids. However, this declaration will be verifiable.

Environmental Sustain for Future Kids
Vancouver, BC CANADA

Email: info@esfk.org
Web: www.esfk.org

Introduction

This is an International Manual dedicated to 'Environmental Sustain for Future Kids' for a sustainable living education. TTAIN & ESFK improves quality of life and reduces environmental degradation by fostering new consumption patterns and sustainable lifestyles through International Cooperative Extension Service programs at houses, offices, schools and libraries all over the globe.

Climate change is real. Therefore people have the potential to make a difference now and for future generations. This manual provides climate science basics, including the roles that lifestyles and populations play in the climate scenario, the significance of carbon footprints, and an overview of the current climate situation. The manual has been categorized based on humanity's needs starting first with food and ending with tourism. The manual then illustrates the difference between adaptation (taking steps to live with the changes) and mitigation (taking steps to slow the rate of change.)

Adaptation examples include food, energy, transportation, recreation. Mitigation focuses on effectively engaging with local governments, through serving on advisory boards, communicating with public officials, educational institutes, schools, universities, libraries and leading communities towards climate change actions.

One useful way to mitigate climate change is through increasing public knowledge to better understand the impact of the rate of change on plants and animals. This is crucial for preserving species; and for assessing potential insects and disease outbreaks in agriculture, natural resources and public health.

Taking personal action is a key element of this manual.

Citizens are challenged to consume 20% fewer resources, to bring world consumption levels down as much as possible. Readers are given 12 practical steps to take to make the changes. The resources section provides additional information, and readers are encouraged to contact the author for further questions.

As an accessibility action, we have provided Online international courses on climate change control as well. You can access the courses via the following link:

http://TopTenAward.org

Step One

Food & Recycling

STEP ONE

Food & Recycling

As an average person drinks 60,000 liters in a lifetime, we are recycling water ourselves. While polluting our oceans we are poisoning our children and grandchildren. Water is natures most vital resource. Let's work together to prevent further pollution of our oceans. Currently, just 25% of mobile phones can be recycled right now. A total of 3.4 million tonnes of plastics were consumed in Australia. A total of 320 000 tonnes of plastics were

recycled, which is an increase of 10 per cent from the 2016-17 recovery. Recycling just one ton of aluminum cans conserves more than 152 million Btu, the equivalent of 1,024 gallons of gasoline or 21 barrels of oil consumed. The universal recycling symbol, logo or icon is an internationally recognized symbol used to designate recyclable materials. The recycling symbol is in the public domain and is not a trademark.

Step Up: Recycle, Reuse & Reduce
How can I recycle?

STEP 1: Go Green & Recycle
Separate recyclables from your trash with a recycling bin. Such as a dual trash can with two compartments or a compost bin. But you can also use a paper shredder for sorting sensitive paperwork that needs shredding before you recycle it. By using these recycling tools you prevent the loss of recyclable materials.

STEP 2: Go Green & Reuse
Plastic disposables and single-use products are wasteful and not stylish. Be fashionable and more eco-friendly and bring your own beautiful and reusable essentials such as a reusable water bottle, coffee cup or fold-able shopping bag. Make a trendy statement and stop plastic pollution.

STEP 3: Go Green & Reduce
How do you reduce plastics and prevent plastic pollution? The answer is: use more natural resources. This is also known as living a zero waste lifestyle where the mission is to reduce waste as much as possible. Check our zero waste store with eco-friendly paper straws, bamboo toothbrushes, and metal safety razors.

STEP 4: Team Up & Go Green
Start a Green Team in your office or workplace together with your colleagues to educate, inspire, challenge and empower employees about your sustainability goals. Know what you throw away today in the office and think about how you and your colleagues can reduce, reuse and recycle tomorrow.

Why is a green lifestyle important?

Nature can't digest plastics because this material is not biodegradable. We can use much more natural resources that are biodegradable by nature itself. Because not 100% of what we consume will be collected or recycled.

The three arrows of the recycling symbol represent the three main stages of the recycling process: recycling, reusing and reducing. These three chasing arrows are also known as the recycling trilogy, and Together the arrows form a closed loop that symbolizes a circular economy.

What is recycling?

Recycling is the process of collecting and processing materials that would otherwise be thrown away as trash and turning them into new products. Recycling can benefit your community, the economy and the environment.

Is recycling truly beneficial for the environment?

The answer is Yes, For example:

Recycling 500 kg of paper can save the energy equivalent of consuming 160 gallons of gasoline.

Recycling just 1000 kg of aluminum cans conserves 1,000 gallons of gasoline or more than 20 barrels of oil consumed.

Plastic bottles are the most recycled plastic product in the United States as of 2015, according to our most recent report. Recycling just 10 plastic bottles saves enough energy to power a laptop for more than 25 hours.

Why is it important to only put items that can be recycled in the recycling bin?

Putting items in the recycling bin that can't be recycled can contaminate the recycling stream. After these unrecyclable items arrive at recycling centers, they can cause costly damage to the equipment. Additionally, after arriving at recycling centers, they must be sorted out and then sent to landfills, which raises costs for the facility. That is why it is important to check with your local recycling provider to ensure that they will accept certain items before placing them into a bin. Some items may also be accepted at retail locations or other at local recycling centers.

INTERNATIONAL MANUAL OF CLIMATE CHANGE CONTROL (IMCCC)• 19

Why are some items that look recyclable not accepted at my recycling facility?

Your local recycling facility might not accept all recyclable items. This is especially true with plastics. While plastic bottles are the most commonly recycled plastic products, other plastics may or may not be accepted in your area, so first check what your local recycling provider accepts. It is important to understand that the existence of a plastic resin code on the product does not guarantee that the product is recyclable in your area. Additionally, glass may not be accepted in some areas, so please confirm with your local provider.

What should I never put in my recycling bin(s)?
Garden hoses
Sewing needles
Bowling balls
Food or food-soiled paper
Propane tanks or cylinders
Aerosol cans that aren't empty

Many communities have collection programs for household hazardous waste to reduce the potential harm posed by these chemicals.

What are the most common items that I can put into my curbside recycling bin?

Cardboard
Paper
Food boxes
Mail
Beverage cans
Food cans
Glass bottles
Jars (glass and plastic)
Jugs
Plastic bottles and caps

Generally, these are the most commonly recycled items. Please confirm with your local recycling provider first before putting these items in your curbside recycling bin, however, since what is accepted depends on your area.

Are paper or plastic shopping bags better for the environment? How about reusable bags versus disposal bags?
We do not have information on the environmental benefits of paper versus plastic bags. We encourages consumers to:

Reduce the number of bags they use,
Reduce the number of bags they throw away after one use,
Reuse bags, and
Recycle bags when they can no longer be used.
Consumers also can reduce waste by using reusable shopping bags.

INTERNATIONAL MANUAL OF CLIMATE CHANGE CONTROL (IMCCC)• 21

Glass
This symbol asks that you recycle the glass container. Please dispose of glass bottles and jars in a bottle bank, remembering to separate colours, or use your glass household recycling collection if you have one.

WHY CAN'T I RECYCLE SOME GLASS ITEMS?
Some types of glass do not melt at the same temperature as bottles and jars. If they enter the glass recycling process it can result in new containers being rejected. These items should be recycled separately - check with your local household waste recycling centre.

HOW TO RECYCLE GLASS BOTTLES AND JARS
Put lids and caps back on. This reduces the chance of them getting lost during the sorting process as they can be recycled separately.
Empty and rinse - a quick rinse will do. Leftover liquid can contaminate other recyclables which may mean they aren't recycled.

Aluminium
This symbol indicates that the item is made from recyclable aluminium.

HOW TO RECYCLE
Rinse or wipe off any crumbs or food residue from foil trays. To rinse just dunk the tray in the washing up water - no need to run the tap.
Scrunch kitchen foil, tub and pot lids and wrappers together to form a ball - the bigger the ball, the easier it is to recycle.
As well as foil, you can usually recycle these other aluminium items:

Drinks cans
Screw top lids from bottles
(recycle with the bottle - the cap can be left on)
Takeaway containers and barbeque trays.

Symbol#1: PET or PETE

PET or PETE (polyethylene terephthalate) is the most common plastic for single-use bottled beverages, because it›s inexpensive, lightweight, and easy to recycle. It poses low risk of leaching breakdown products. Its recycling rates remain relatively low (around 20%), even though the material is in high demand by manufacturers.

Found in:
Soft drinks, water, ketchup mouthwash bottles; peanut butter containers; salad dressing and vegetable oil containers

How to recycle it:
PET or PETE can be picked up through most curbside recycling programs as long as it›s been emptied and rinsed of any food. When it comes to caps, our environmental pros say it's probably better to dispose of them in the trash (since they›re usually made of a different type of plastic), unless your town explicitly says you can throw them in the recycle bin. There›s no need to remove bottle labels because the recycling process separates them.

Recycled into:
Polar fleece, fiber, tote bags, furniture, carpet, paneling, straps, bottles and food containers (as long as the plastic being recycled meets purity standards and doesn›t have hazardous contaminants)

Symbol#2: HDPE

HDPE (high density polyethylene) is a versatile plastic with many uses, especially when it comes to packaging. It carries low risk of leaching and is readily recyclable into many types of goods.

Found in: Milk jugs; juice bottles; bleach, detergent, and other household cleaner bottles; shampoo bottles; some trash and shopping bags; motor oil bottles; butter and yogurt tubs; cereal box liners

How to recycle it: HDPE can often be picked up through most curbside recycling programs, although some allow only containers with necks. Flimsy plastics (like grocery bags and plastic wrap) usually can't be recycled, but some stores will collect and recycle them.

Recycled into: Laundry detergent bottles, oil bottles, pens, recycling containers, floor tile, drainage pipe, lumber, benches, doghouses, picnic tables, fencing, shampoo bottles

Symbol#3: PVC or V

PVC (polyvinyl chloride) and V (vinyl) is tough and weathers well, so it's commonly used for things like piping and siding. PVC is also cheap, so it's found in plenty of products and packaging. Because chlorine is part of PVC, it can result in the release of highly dangerous dioxins during manufacturing. Remember to never burn PVC, because it releases toxins.

Found in: Shampoo and cooking oil bottles, blister packaging, wire jacketing, siding, windows, piping

How to recycle it: PVC and V can rarely be recycled, but it's accepted by some plastic lumber makers. If you need to dispose of either material, ask your local waste management to see if you should put it in the trash or drop it off at a collection center.

Recycled into: Decks, paneling, mud-flaps, roadway gutters, flooring, cables, speed bumps, mats

Symbol#4: LDPE

LDPE (low density polyethylene) is a flexible plastic with many applications. Historically, it hasn't been accepted through most American recycling programs, but more and more communities are starting to accept it.

Found in: Squeezable bottles; bread, frozen food, dry cleaning, and shopping bags; tote bags; furniture

How to recycle it: LDPE is not often recycled through curbside programs, but some communities might accept it. That means anything made with LDPE (like toothpaste tubes) can be thrown in the trash. Just like we mentioned under HDPE, plastic shopping bags can often be returned to stores for recycling.

Recycled into: Trash can liners and cans, compost bins, shipping envelopes, paneling, lumber, landscaping ties, floor tile

Symbol#5: PP

PP (polypropylene) has a high melting point, so it's often chosen for containers that will hold hot liquid. It's gradually becoming more accepted by recyclers.

Found in: Some yogurt containers, syrup and medicine bottles, caps, straws

How to recycle it: PP can be recycled through some curbside programs, just don't forget to make sure there's no food left inside. It's best to throw loose caps into the garbage since they easily slip through screens during recycling and end up as trash anyways.

Recycled into: Signal lights, battery cables, brooms, brushes, auto battery cases, ice scrapers, landscape borders, bicycle racks, rakes, bins, pallets, trays

Symbol#6: PS

PS (polystyrene) can be made into rigid or foam products — in the latter case it is popularly known as the trademark Styrofoam. Styrene monomer (a type of molecule) can leach into foods and is a possible human carcinogen, while styrene oxide is classified as a probable carcinogen. The material was long on environmentalists' hit lists for dispersing widely across the landscape, and for being notoriously difficult to recycle. Most places still don't accept it in foam forms because it's 98% air.

Found in: Disposable plates and cups, meat trays, egg cartons, carry-out containers, aspirin bottles, compact disc cases

How to recycle it: Not many curbside recycling programs accept PS in the form of rigid plastics (and many manufacturers have switched to using PET instead). Since foam products tend to break apart into smaller pieces, you should place them in a bag, squeeze out the air, and tie it up before putting it in the trash to prevent pellets from dispersing.

Recycled into: Insulation, light switch plates, egg cartons, vents, rulers, foam packing, carry-out containers

Symbol#7: MISCELLANEOUS

A wide variety of plastic resins that don't fit into the previous categories are lumped into this one. Polycarbonate is number seven plastic, and it's the hard plastic that has worried parents after studies have shown it as a hormone disruptor. PLA (polylactic acid), which is made from plants and is carbon neutral, also falls into this category.

Found in: Three- and five-gallon water bottles, bullet-proof materials, sunglasses, DVDs, iPod and computer cases, signs and displays, certain food containers, nylon

How to recycle it: These other plastics are traditionally not recycled, so don't expect your local provider to accept them. The best option is to consult your municipality's website for specific instructions.

Recycled into: Plastic lumber and custom-made products

Compostable

Compostable
Products certified to be industrially compostable according to the European standard EN 13432/14955 may bear the 'seedling' logo.

Never place compostable plastic into the recycling with other plastics; as it is designed to break down it cannot be recycled and contaminates recyclable plastics. Plastics that carry this symbol can be recycled with your garden waste through your local authority.

Waste electricals
This symbol explains that you should not place the electrical item in the general waste. Electrical items can be recycled through a number of channels.

Step Two

Energy

STEP TWO

What is Renewable energy ?

Renewable energy is energy that has been derived from earth's natural resources that are not finite or exhaustible, such as wind and sunlight. Renewable energy is an alternative to the traditional energy that relies on fossil fuels, and it tends to be much less harmful to the environment.

Types of Renewable Energy :

Solar energy is derived by capturing radiant energy from sunlight and converting it into heat, electricity, or hot water. Photovoltaic (PV) systems can convert direct sunlight into electricity through the use of solar cells.

Wind farms capture the energy of wind flow by using turbines and converting it into electricity. There are several forms of systems used to convert wind energy and each vary. Commercial grade wind-powered generating systems can power many different organizations, while single-wind turbines are used to help supplement pre-existing energy organizations.

Hydroelectric, Dams are what people most associate when it comes to hydroelectric power. Water flows through the dam's turbines to produce electricity, known as pumped-storage hydropower. Run-of-river hydropower uses a channel to funnel water through rather than powering it through a dam.

Geothermal heat is heat that is trapped beneath the earth's crust from the formation of the Earth 4.5 billion years ago and from radioactive decay. Sometimes large amounts of this heat escapes naturally, but all at once, resulting in familiar occurrences, such as volcanic eruptions and geysers.

Ocean, the ocean can produce two types of energy: thermal and mechanical. Ocean thermal energy relies on warm water surface temperatures to generate energy through a variety of different systems. Ocean mechanical energy uses the ebbs and flows of the tides to generate energy, which is created by the earth's rotation and gravity from the moon.

Hydrogen needs to be combined with other elements, such as oxygen to make water as it does not occur naturally as a gas on its own. When hydrogen is separated from another element it can be used for both fuel and electricity.

Bioenergy is a renewable energy derived from biomass. Biomass is organic matter that comes from recently living plants and organisms. Using wood in your fireplace is an example of biomass that most people are familiar with.

Renewable Energy: What Can You Do?

As a consumer you have several opportunities to make an impact on improving the environment through the choice of a greener energy solution. If you're a homeowner, you have the option of installing solar panels in your home. Solar panels not only reduce your energy costs, but help improve your standard of living with a safer, more eco-friendlier energy choice that doesn't depend on resources that harm the environment. There are also alternatives for a greener way of life offered by your electric companies. Just Energy allows consumers to choose green energy options that help you reduce your footprint with energy offsets.

Solar

Solar energy is derived by capturing radiant energy from sunlight and converting it into heat, electricity, or hot water. Photovoltaic (PV) systems can convert direct sunlight into electricity through the use of solar cells.

Benefits

One of the benefits of solar energy is that sunlight is functionally endless. With the technology to harvest it, there is a limitless supply of solar energy, meaning it could render fossil fuels obsolete. Relying on solar energy rather than fossil fuels also helps us improve public health and environmental conditions. In the long term, solar energy could also eliminate energy costs, and in the short term, reduce your energy bills. Many federal local, state, and federal governments also incentivize the investment in solar energy by providing rebates or tax credits.

Current Limitations

Although solar energy will save you money in the long run, it tends to be a significant upfront cost and is an unrealistic expenses for most households. For personal homes, homeowners also need to have the ample sunlight and space to arrange their solar panels, which limits who can realistically adopt this technology at the individual level.

Wind

Wind farms capture the energy of wind flow by using turbines and converting it into electricity. There are several forms of systems used to convert wind energy and each vary. Commercial grade wind-powered generating systems can power many different organizations, while single-wind turbines are used to help supplement pre-existing energy organizations. Another form is utility-scale wind farms, which are purchased by contract or wholesale. Technically, wind energy is a form of solar energy. The phenomenon we call "wind" is caused by the differences in temperature in the atmosphere combined with the rotation of Earth and the geography of the planet.

Benefits

Wind energy is a clean energy source, which means that it doesn't pollute the air like other forms of energy. Wind energy doesn't produce carbon dioxide, or release any harmful products that can cause environmental degradation or negatively affect human health like smog, acid rain, or other heat-trapping gases. Investment in wind energy technology can also open up new avenues for jobs and job training, as the turbines on farms need to be serviced and maintained to keep running.

Current Limitations

Since wind farms tend to be built in rural or remote areas, they are usually far from bustling cities where the electricity is needed most. Wind energy must be transported via transition lines, leading to higher costs. Although wind turbines produce very little pollution, some cities oppose them since they dominate skylines and generate noise. Wind turbines also threaten local wildlife like birds, which are sometimes killed by striking the arms of the turbine while flying.

Hydroelectric

Dams are what people most associate when it comes to hydroelectric power. Water flows through the dam's turbines to produce electricity, known as pumped-storage hydro power. Run-of-river hydro power uses a channel to funnel water through rather than powering it through a dam.

Benefits

Hydroelectric power is very versatile and can be generated using both large scale projects, like the Hoover Dam, and small scale projects like underwater turbines and lower dams on small rivers and streams. Hydroelectric power does not generate pollution, and therefore is a much more environmentally-friendly energy option for our environment.

Current Limitations

Most hydroelectricity facilities use more energy than they are able to produce for consumption. The storage systems may need to use fossil fuel to pump water. Although hydroelectric power does not pollute the air, it disrupts waterways and negatively affects the animals that live in them, changing water levels, currents, and migration paths for many fish and other freshwater ecosystems.

Geothermal

Geothermal heat is heat that is trapped beneath the earth's crust from the formation of the Earth 4.5 billion years ago and from radioactive decay. Sometimes large amounts of this heat escapes naturally, but all at once, resulting in familiar occurrences, such as volcanic eruptions and geysers. This heat can be captured and used to produce geothermal energy by using steam that comes from the heated water pumping below the surface, which then rises to the top and can be used to operate a turbine.

Benefits

Geothermal energy is not as common as other types of renewable energy sources, but it has a significant potential for energy supply. Since it can be built underground, it leaves very little footprint on land. Geothermal energy is naturally replenished and therefore does not run a risk of depleting (on a human timescale).

Current Limitations

Cost plays a major factor when it comes to disadvantages of geothermal energy. Not only is it costly to build the infrastructure, but another major concern is its vulnerability to earthquakes in certain regions of the world.

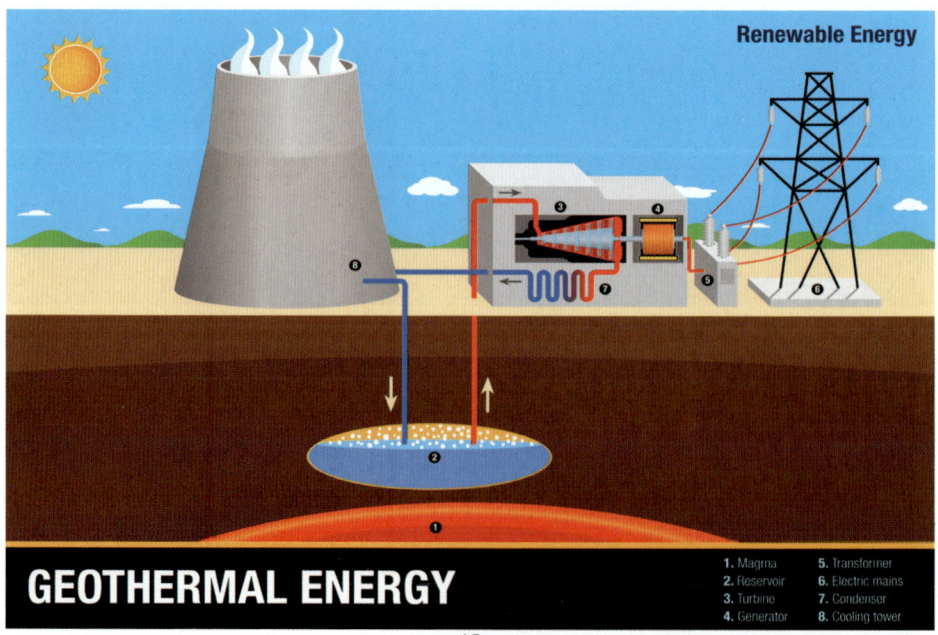

Ocean

The ocean can produce two types of energy: thermal and mechanical. Ocean thermal energy relies on warm water surface temperatures to generate energy through a variety of different systems. Ocean mechanical energy uses the ebbs and flows of the tides to generate energy, which is created by the earth's rotation and gravity from the moon.

Benefits

Unlike other forms of renewable energy, wave energy is predictable and it's easy to estimate the amount of energy that will be produced. Instead of relying on varying factors, such as sun and wind, wave energy is much more consistent. This type of renewable energy is also abundant, the most populated cities tend to be near oceans and harbors, making it easier to harness this energy for the local population. The potential of wave energy is an astounding as yet untapped energy resource with an estimated ability to produce 2640 TWh/yr. Just 1 TWh/yr of energy can power around 93,850 average U.S. homes with power annually, or about twice than the number of homes that currently exist in the U.S. at present.

Current Limitations

Those who live near the ocean definitely benefit from wave energy, but those who live in landlocked states won't have ready access to this energy. Another disadvantage to ocean energy is that it can disturb the ocean's many delicate ecosystems. Although it is a very clean source of energy, large machinery needs to be built nearby to help capture this form energy, which can cause disruptions to the ocean floor and the sea life that habitats it. Another factor to consider is weather, when rough weather occurs it changes the consistency of the waves, thus producing lower energy output when compared to normal waves without stormy weather.

Hydrogen
Hydrogen needs to be combined with other elements, such as oxygen to make water as it does not occur naturally as a gas on its own. When hydrogen is separated from another element it can be used for both fuel and electricity.

Benefits
Hydrogen can be used as a clean burning fuel, which leads to less pollution and a cleaner environment. It can also be used for fuel cells which are similar to batteries and can be used for powering an electric motor.

Current Limitations
Since hydrogen needs energy to be produced, it is inefficient when it comes to preventing pollution.

Biomass

Bioenergy is a renewable energy derived from biomass. Biomass is organic matter that comes from recently living plants and organisms. Using wood in your fireplace is an example of biomass that most people are familiar with.

There are various methods used to generate energy through the use of biomass. This can be done by burning biomass, or harnessing methane gas which is produced by the natural decomposition of organic materials in ponds or even landfills.

Benefits

The use of biomass in energy production creates carbon dioxide that is put into the air, but the regeneration of plants consumes the same amount of carbon dioxide, which is said to create a balanced atmosphere. Biomass can be used in a number of different ways in our daily lives, not only for personal use, but businesses as well. In 2017, energy from biomass made up about 5% of the total energy used in the U.S. This energy came from wood, biofuels like ethanol, and energy generated from methane captured from landfills or by burning municipal waste.

Current Limitations

Although new plants need carbon dioxide to grow, plants take time to grow. We also don't yet have widespread technology that can use biomass in lieu of fossil fuels.

The Future of Renewable energy

Take a moment to close your eyes and imagine the world ten, twenty, fifty years from now. How do you heat your home? What do our energy systems look like? How about our cars – how are they fueled?

In an ideal world, renewable energy will become the primary source of the planet's energy, as opposed to traditional energy sources, like fossil fuels (which release harmful carbon emissions and pollution into the atmosphere). So, what does the future of renewable energy actually look like? Time will tell – but these crazy, cool, new innovations may provide a glimpse into the future of renewables:

Solar Powered Panels that Chase the Sun

Genius in its simplicity, this new technology overcomes one of the biggest challenges facing solar power – clouds and inclement weather. These solar panels actually reposition themselves to soak in the most possible sunlight, resulting in much higher levels of efficiency.

Solar/Wind Hybrids

As solar and wind technologies continue to improve, scientists and engineers are experimenting with ways to make both more efficient. Bring on the superhero of renewable energy: solar and wind hybrids. This technology combines wind turbines with solar photovoltaic (PV) panels to produce higher levels of energy – and studies have found that they are nearly twice as efficient.

Energy From Unusual Sources

You've heard about energy from the wind, from the sun, and even from compost or other organic sources, but how about algae? It's true – "algae energy" is a concept that scientists are currently developing. Another awesome technology: these batteries made from wood (oh hey there, bioenergy). Color us impressed.

Do-It-Yourself Renewable Energy

We dream of a world with solar panels on every roof, wind turbines in every backyard. Is this a realistic dream? Scientists and engineers are getting closer every day. Even today, some dedicated homeowners have taken pains to install their own personal systems of solar power to heat/power their homes – a trend we hope to see continue well into the future.

Nuclear Energy Is Extraordinary

Nuclear energy comes from splitting atoms in a reactor to heat water into steam, turn a turbine and generate electricity. Ninety-four nuclear reactors in 28 states generate nearly 20 percent of the nation's electricity, all without carbon emissions because reactors use uranium, not fossil fuels. These plants are always on: well-operated to avoid interruptions and built to withstand extreme weather, supporting the grid 24/7.

Step Three

Fashion & Textile

STEP THREE

All about 'Eco-friendly' fashion and textile

ECO-FRIENDLY FASHION

ORGANIC COTTON CULTIVATION

FABRICS FROM EASILY RENEWABLE CROPS

NATURAL DYES FROM PLANTS

JEANS MADE OF REPURPOSED DENIM

The textile industry being a very good example for the most advancing and ecologically harmful industry in the world, various innovations are done in order to safeguard our mother earth. The production stages of textile include bleaching, dyeing etc...Contribute to a large extend of pollution thus making it important to make it more sustainable. Controlling pollution is as vital as making a product free from the toxic effect.So in order to safeguard our environment we must take some preventive measures and technologies that can maintain the balance of our eco system and makes the final product free from toxic effects. Generally there is really no such thing as a 100% eco friendly piece of clothing because all clothing takes water (for the fibres to grow) and energy (to make the fabric and the final garments).So, Eco-friendly clothing can be termed as a clothing made of natural fibres such as organic cotton and hemp, clothing that has been organically dyed with vegetables or any fabrics that use small amounts of water, energy and chemicals that affect the environment.

Natural fibres have intrinsic properties such as mechanical strength, low weight and healthier to the wearer that has made them particularly attractive. The word 'eco' is short for ecology. Ecology is the study of the interactions between organisms and their environment. Therefore 'eco' friendly (or 'ecology friendly') is a term to refer to goods and services considered to inflict minimal or no harm on the environment. "Think globally, act locally" is the slogan of tomorrow for the world textile industry.

ECO-FRIENDLY Textiles

Any textile product, which is produced in eco-friendly manner and processed under eco-friendly limits, is known as eco friendly textiles. It is also known as **sustainable fashion**, **eco fashion** and **Ecotech**. Materials can be considered as "Eco-friendly" on the basis of various factors:

- Renewability of the product
- Ecological footprint of resources - how much land it takes for the full growth of a product
- Determining the eco friendliness of a product - amount of chemicals required for the production of products.

TTAIN Eco-Review

We have created our own Fibre Eco-Review, using different resources and studies on the environmental impact of each of the fibres. Here, we have focused on the fibre production.

We have divided fibres in two main categories according to their environmental impact:

- **TTAIN Recomended (TTAIN-R)**
- **TTAIN Not Recommended (TTAIN-NR)**

TTAIN Recomended (TTAIN-R) Fibres

Group A Plant Based	Group B Animal based	Group C Recycled	Group D Semi-Synthetic
Linen	Silk	Recycled Polyester	Pineapple, Corn, Milk, Banana, Orange
Organic Cotton	Alpaca	Recycled Nylon	Lycocell/ Tencel
Hemp	Sustainable Wool	Recycled Cotton	Algae Fibres
Ramie	Sustainable Cashmere	Recycled Wool	Cupro
Natural Rubber	Sustainable Leather	Recycled Textile Fibres	Ayurvastra
Jute	Responsible Down		Soya Bean

TTAIN Not Recomended (TTAIN-NR) Fibres

Group E Natural and Animal Based Fibres		Group F Synthetic and Semi-Synthetic Fibres	
Wool	Leather	Polyester	Acrylic
Cotton	Cashmere	Rayon	Bamboo
Down		Viscose	Vegan Leather
		Modal	Nylon
		Spandex	Polypropylene
		Aramide	PVC

Group A: Plant Based

Linen

Linen is a natural fibre which stems from the flax plant. It uses considerably fewer resources than cotton or polyester (such as water, energy, pesticides, insecticides, fertilizers).

Flax can grow in poor soil which is not used for food production. In some cases, it can even rehabilitate polluted soil. Flax plants also have a high rate of carbon absorption.

Organic Cotton

The fabric has the same quality as conventional cotton but not the negative impact on the environment. Organic cotton addresses most of the environmental challenges which conventional cotton production faces. It is grown from non-GMO seeds and without the use of pesticide, insecticide or fertilizer. Unlike conventional cotton, organic farmers use ancestral farming methods, including crop-rotation, mixed farming or no-till farming to pre-

serve the soil. Organic cotton uses up to 67% less water than conventional cotton according to some sources. Organic cotton farmers are not exposed to harmful substances. Several organizations have established certifications for organic cotton such as GOTS. Certification is the only proof that a product is truly organic.

Hemp

Hemp fabric comes from the plant with the same name. It is one of the fastest growing plants and it doesn't need much water, energy, pesticide, or fertilizers. The plant is very good for soil, it can be grown for many years in the same place without exhausting it. This is why hemp is considered to be eco-friendly. Hemp has very similar properties to linen. They are often difficult to differentiate. However, as hemp belongs to the same family as cannabis (although it does not have the same psychoactive effects), growing hemp is heavily regulated or prohibited in many countries.

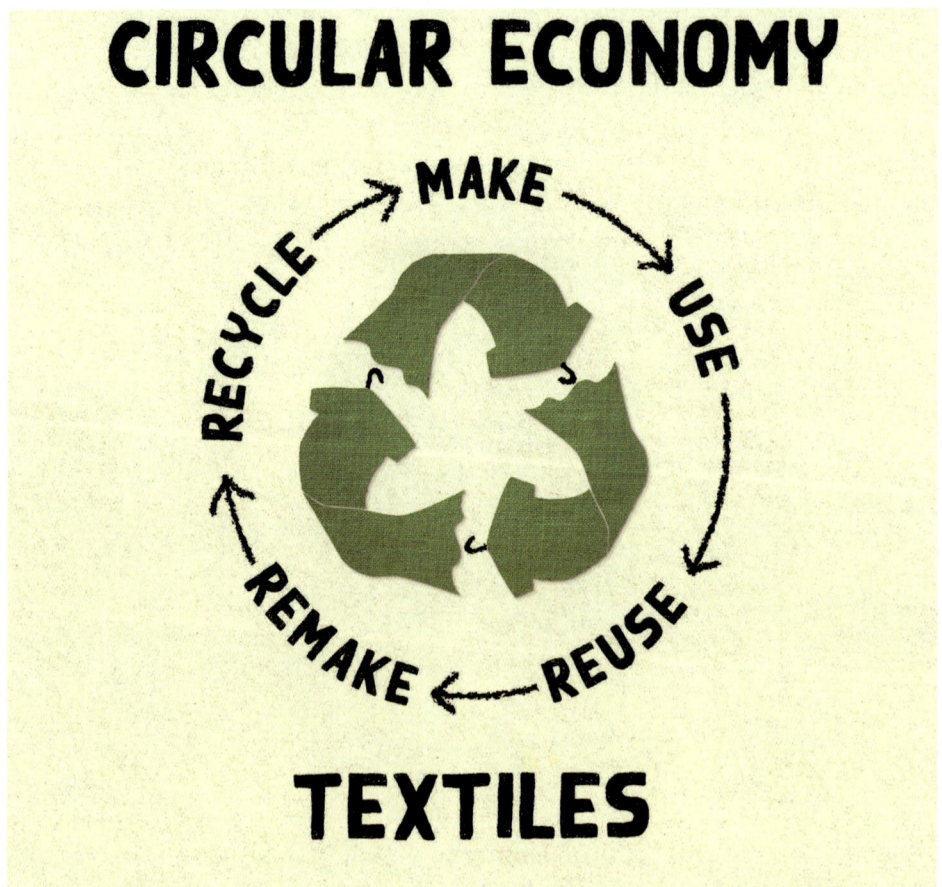

Ramie

Ramie and stinging nettle, or European nettle, are plants used to produced a fibre similar to linen. They are not very common but they are considered sustainable.

Natural Rubber

Synthetic rubber is basically plastic whereas natural rubber is made from the milk of the Hevea tree. Most of the soles of our shoes are nowadays made with synthetic rubber which is a very different thing from natural rubber. Natural rubber, therefore, comes from a renewable resource, the harvesting of rubber doesn't harm trees but actually helps the tree to flourish. It protects forests from being cut down as it gives value to the exploitation of the tree. Rubber is also easy to recycle & biodegradable. Rubber from FSC®-certified forest (IEL, Vol.1, page: 39) is even better as it ensures the good environmental management of the forest.

Jute

Jute fibre is 100% bio-degradable and recyclable and thus environmentally friendly. Jute, an edible leafy vegetable, also known as "the golden fibre", is a long, soft and shiny fibre made from the cellulose and lignin material from the jute plant. A hectare of jute plants consumes about 14.5 tonnes of carbon dioxide and releases 10.5 tonnes of oxygen. Jute also does not generate toxic gases when burnt. Jute reaches maturity quickly, between 5-7 months, making it an incredibly efficient source of renewable material, and therefore "sustainable". Jute products help in decreasing environmental pollution as its use decreases the demand for plastic bags which are non-bio degradable and pollute the surroundings. Jute bags are more useful as compared to the plastic bags as they can be used again and again. **you can simply use jute bags in lieu of non-biodegradable plastic**. Jute is also compostable by itself just like egg shells or the melon peels which means that you can sleep easy knowing that you are not contributing to the pollution or harmful clogging of our environment.

Group B: Animal Based

Silk

Silk is a protein fibre spun by silkworms and is a renewable resource. Silk is also biodegradable. For these reasons, we consider silk a sustainable fibre. However, chemicals are used to produce conventional silk, so we will always consider organic silk to be a better option. Because conventional silk production kills the silkworm, animal rights advocates prefer "Peace Silk", Tussah, Ahimsa silks which allow the moth to evacuate the cocoon before it is boiled to produce silk.

Alpaca

Alpaca fibre comes from the fleece of the animal bearing the same name. Alpacas are mainly bred in the Peruvian Andes. Alpacas are much more eco-friendly than cashmere goats, because they cut the grass they eat instead of pulling it out, which allows for the grass to keep growing. Additionally, Alpacas have soft padding under their feet, which is more gentle for the soil than goat or sheep hooves.

They need very little water and food to survive and produce enough wool for 4 or 5 sweaters per year while a goat needs 4 years to produce just one cashmere sweater.

Sustainable Wool

Conventional wool is far from being as eco-friendly as we would expect. However, there are some sustainable wool options on the market which make it possible for us to dress warmly and sustainably. So far, we have found the Responsible Wool Standard (RWS), which ensures that farms use best practices to protect the land, and treat the animal decently. Certified organic wool guarantees that pesticides and parasiticides are not used on the pastureland or on the sheep themselves, and that good cultural and management practices of livestock are used. Certified organic wool is still pretty rare on the market. GOTS seems to be the only organization certifying organic wool.

Sustainable Cashmere

As we can see in the related section, conventional cashmere has very significant consequences for the environment.

The good news is that there are a few sustainable cashmere options which address these environmental problems and give us the possibility to buy cashmere without a guilty conscience.

Sustainable Leather

Leather will never be an animal-friendly product: It is made of dead animal skin. However, the skin used to make leather comes from animals raised for their meat. In that sense, it uses a byproduct from another industry, so it doesn't actually need additional land and resources. Conventional leather is heavily criticized for the environmental impact of the tanning process. But leather can also be eco-friendly. There are not many options in the market yet, but they do exist.

Responsible Down

The main issue of conventional down is the live-plucking of birds which is cruel and painful to the animal. For those wanting to use down and enjoy its durability, its lightweight, and warmth, we recommend looking for certified responsible down (Responsible Down Standard) or recycled down.

INTERNATIONAL MANUAL OF CLIMATE CHANGE CONTROL (IMCCC) • 63

Group C: Recycled

Recycled polyester

Recycled polyester, often called rPet, is made from recycled plastic bottles. It is a great way to divert plastic from our landfills. The production of recycled polyester requires far fewer resources than that of new fibres and generates fewer CO_2 emissions.

There are 2 ways to recycle polyester: For mechanical recycling, plastic is melted to make new yarn. This process can only be done a few times before the fibre loses its quality. Chemical recycling involves breaking down the plastic molecules and reforming them into yarn. This process maintains the quality of the original fibre and allows the material to be recycled infinitely, but it is more expensive.

Although, Recycled polyester is definitely a sustainable option but, we need to be aware that it is still non-biodegradable and takes years to disappear once thrown away.

Recycled Nylon

Recycled Nylon has the same benefits as recycled polyester: It diverts waste from landfills and its production uses much fewer resources than virgin nylon (including water, energy and fossil fuel).

A large part of the recycled nylon produced comes from old fishing nets. This is a great solution to divert garbage from the ocean. It also comes from nylon carpets, tights, etc. Recycling nylon is still more expensive than new nylon, but it has many environmental advantages. A lot of research is currently being conducted to improve the quality and reduce the costs of the recycling process.

Recycled Cotton

Recycled cotton prevents additional textile waste and requires far fewer resources than conventional or organic cotton. This makes it a great sustainable option. Cotton can be recycled using old garments or textile leftovers. The quality of the cotton may be lower than of new cotton. Recycled cotton is therefore usually blended with new cotton. The production of recycled cotton is still very limited.

Recycled Wool

Recycled wool is also very sustainable option. Apart from diverting used wool garments from landfills, it saves a considerable amount of water, reduces land use for sheep grazing and avoids the use of chemicals for dyeing. Recycled wool contributes to a reduction of air, water, and soil pollution.

Recycled Textile Fibres

A lot of researches are currently going in that direction: making textile from textile waste. As we generate so much textile pre-consumer as well as post-consumer waste, it makes total sense to re-use it instead of throwing it away. However, due to the difficulty to separate fibres blend and other technological challenges, this type of textile is not yet easily available.

Group D

Group D: Semi-Synthetic
Pineapple, Corn, Milk, Bannana and Orange Fibres

Pineapple Fibre (Piñatex)
Piñatex is a fibre that comes from pineapple leaves. It is considered sustainable because it uses the by-products of pineapple harvests, so there is no need for extra resources to produce it. It is used as a substitute for leather.

Corn Fibre
Corn is available in both spun and filament forms. It is derived from naturally occurring plant sugars. It balances strength and resilience with comfort, softness and drape in textiles. Corn also uses no chemical additives or surface treatments and is naturally flame retardant. Corn fibre manufacturers have claimed that these fibres can be used for sportswear, jacket, outer coat, apparels etc.

Milk Fibre
Milk Fibre was firstly introduced in 1930 in Italy & America to compete the wool. It is the new innovative Fibre and a kind of synthetic Fibre made of milk casein Fibre through bio-engineering method. It can also be used to create top-grade underwear, shirts, T-shirts, loungewear, etc. It contains seventeen amino acids & natural anti-bacterial rate is above eighty percent. Hence milk fibre has sanitarian function.

Banana Fibre
The use of banana stems as a source of fibre such as cotton and silk is becoming popular now. It is used all over the world for multiple purposes such as making tea bags or sanitary napkins to Japanese yen notes and car tyres. It is also known as musa fibre which is one of the strongest natural fibres. Banana stem, hitherto considered a complete waste, is now being made into banana-fibre cloth which comes in differing weights and thicknesses based on what part of the banana stem the fibre was taken from. The innermost sheaths are where the softest fibres are obtained, and the thicker and sturdier fibres come from the outer sheaths. High water absorbing property of this fabric makes this clothing cool to wear.

Orange Fibre
Orange Fibre is an innovative fabric made from orange skins that comes from the juice industry wastes.

Lyocell (Tencel)
Lyocell is made in a closed-loop system that recycles almost all of the chemicals used. Lyocell is a manufacturing process of rayon which is much more eco-friendly than its relatives modal and viscose. "Lyocell" is the generic name of the manufacturing process and fibre. Tencel is the brand name of the lyocell commercialized by the company Lenzing AG. Just like rayon and viscose, lyocell is more than 95% biodegradable.

Algae Fibres
Algae are being tapped as a new resource to make fibres, finishes and dyes for the textile industry. Algae bloom can provide cellulose or proteins, and in microalgae form, the species can produce non-petrochemical oils.

The food, pharmaceuticals and biofuels industries have been harnessing the ability of microalgae to produce compounds on an industrial scale for years. Now, a new generation of companies is set on putting this quality to use in developing materials and supplies for textiles, apparel and footwear. Many of the projects involving algae and textiles are still in research phase, but the landscape is changing fast.

Cupro
Cupro is an artificial cellulose fibre made from Linter Cotton (or Cotton wastes). In order to obtain the ready to weave yarn, the extracted cellulose is soaked in a bath of a chemical solution called «cuprammonium », hence the Cupro Name. All the process is made in closed-loop. The large quantities of water and chemicals used in the production of Cupro are therefore constantly reused until they are completely exhausted. The chemicals used are free of toxic or dangerous compounds for health and the environment. Cupro is also biodegradable, so it considers a good eco-friendly alternative to viscose.

Ayurvastra

Ayur vastra is a Sanskrit term made up of two words "AYUR" means "health" & "VASTRA" means "Cloth", meaning "life cloth". It is a branch of Ayurveda. Ayur vastra cloth is completely free from synthetic chemicals & toxic substances making this cloth organic, sustainable & biodegradable. Ayur vastra or medical dress is made of 100% pure organic cotton or silk, wool, jute & coir products that have been hand loomed, dyed by using various Ayurveda herbs & have medicinal qualities. Herb dyed organic fabrics act as healing agents or as an absorber through skin. Each fabric is infused with specific herbs that can help treat skin conditions. Herbs used in Ayur vastra are known to cure allergies having anti-microbial, anti-inflammatory properties Ayur vastra is extra smooth & good for transpiration that helps in recovering various diseases. It may help treat a broad range of diseases such as skin infections, diabetes, eczema, psoriasis, hypertension, high blood pressure, asthma & insomnia.

Soya Bean Fibre

Soybean fibre is a sustainable textile fibre made from renewable natural resources. The soybean protein fibre is actually made from the byproduct leftovers of soybean oil/tofu/soymilk production, which would normally be discarded.

Cotton

Cotton is mainly produced in dry and warm regions, but it needs a lot of water to grow. In some places, like India, inefficient water use means that more than 19,000 liters of water are needed to produce 1000g of cotton. In the meantime, 100 million people in India do not have access to drinking water. Although it is a natural fibre, conventional cotton is far from environmentally friendly. 97% of cotton is grown using fertilizers and genetically modified seeds. Cotton represents 11% of the pesticides and 27% of the insecticides used globally. 95% of the world's cotton farmers are located in developing countries where labor, health and safety regulations are nonexistent or not enforced most of the time. Child and forced labor are common practice. In some countries, people are forced to pick cotton for little or no pay every year.

Wool

Wool as such is a renewable natural fibre, so it could have been considered an environment-friendly option. Unfortunately, the extensive sheep farming practiced to meet the global demands has had disastrous consequences on the environment. Sheep survive by grazing, which can have a positive impact on certain types of ecosystems when it is well managed. But when the land is grazed too heavily, this leads to overgrazing. Overgrazing means that the vegetation does not have enough time to grow back before it is consumed. The soil becomes weak and vulnerable to erosion and desertification.

For example, 29% of the region of Patagonia is affected by desertification, mainly due to overgrazing by sheep which are primarily raised for their wool. Sheep also release methane, a gas that is 23 times worse for global warming than CO_2. Sheep are often subjected to insecticide baths which contain substances hazardous to the farmers. Residues of those harmful chemicals can remain in the wool and make its way into our clothes. Another concern about wool production is the poor treatment of sheep. When a sheep's fleece is removed (shearing), the shearers often hurt the animals, cutting their skin or hitting them to keep them quiet. Finally, the practice of mulesing has been widely denounced by animal rights activists. Mulesing involves removing the skin of the Merino sheep around the breech to prevent parasitic infection.

Down

Down is the layer of the fine feather of birds. Down has been used for a very long time for insulation and pillows and duvet. It is a light and warm material and very long-lasting. The main sustainability issue with down is that part of the world's supply of down feathers is directly taken ("plucked") on live birds. This practice has been largely denounced due to the suffering of the animal. It is now banned in some countries but still authorized in others. When buying down, it is essential to look for responsible down.

Leather

Leather is a controversial fibre. First of all, it is not an animal-friendly option, since it is made of dead animal skin. But environmental and social concerns related to leather are mostly linked to the tanning process: Toxic chemicals are used (chromium in 79% of cases) to transform the skins into wearable leather. Those substances are often dumped into rivers, polluting freshwater and oceans. Also, most of the tanning factory workers around the world do not wear adequate protection and suffer from skin, eye, and respiratory diseases, cancer and more due to their exposure to chemical substances.

Cashmere

Cashmere fibre comes from cashmere goat hairs. More than 77% of the world's cashmere is produced in China and Mongolia. The main environmental issue stemming from cashmere is due to the fact that goats pull the grass out by the roots when they eat instead of cutting it. As a result, the grass does not grow back, leading to land desertification. This, combined with an overpopulation of goats, results in a real environmental threat.
Mongolia is now suffering the consequences of this overgrazing through cashmere goats. The breeding of more than 21 million cashmere goats is the principal cause of the massive desertification threatening 93% of the surface of the country.

Group F

Polyester

Polyester is the most common fibre in our garment. We can find it in 55% of our clothes. Polyester is a synthetic fibre derived from petroleum, a non-renewable fossil fuel. As we know, the transformation of crude oil into petrochemicals releases toxins into the atmosphere that are dangerous for human and ecosystem health. The production of polyester also highly energy intensive. One of the major problems with this plastic fibre, is the fact that it is non-biodegradable. Furthermore, each time we wash a polyester garment, it releases 700.000 plastic microfibres, ending up in rivers and oceans and then in our food chain.

Viscose, Rayon, Modal

Viscose (also called Artificial Silk or Art Silk) is the most common type of rayon. Viscose production involves a lot of chemicals, heavily harmful to the environment when they are released in effluents. Rayon is a fibre from regenerated cellulose, generally derived from wood pulp. Rayon is usually made from eucalyptus trees, but any plant can be used (such as bamboo, soy, cotton, etc). To produce the fibre, the plant cellulose goes through a process involving a lot of chemicals, energy and water. Solvents used during the process can be very toxic to humans and to the environment. Viscose, modal, lyocell and bamboo are different types of rayon. The other substantial environmental concerns arising from rayon production is the massive deforestation involved. Thousands of hectares of rainforest are cut down each year to plant trees specifically used to make rayon. Only a very small percentage of this wood is obtained through sustainable forestry practices.

Modal, another type of rayon using beech trees with a similar process to viscose. However modal is produced by many other manufacturers who don't necessarily use sustainable processes and it is now rather easy to find sustainable fibres in the market.

Bamboo
Bamboo is usually sold as an eco-friendly textile. Which is partially true, as the bamboo plant is potentially one of the world's most sustainable resource. It grows very quickly and easily, it doesn't need pesticide or fertilizers, and it doesn't need to be replanted after harvest because it grows new sprouts from the roots. However, to turn bamboo into fibre, bamboo is processed with strong chemical solvents that are potentially harmful to the health of manufacturing workers, the consumers wearing the garment, and for the environment when chemicals are released in wastewater.

Acrylic, polyamide, nylon, polypropylene, PVC, spandex
Acrylic, polyamide, nylon, polypropylene, PVC, spandex (AKA lycra or elastane), aramide, etc, are all different types of synthetic fibres that are derived from petroleum and therefore have a very similar impact on the environment as polyester.

Vegan Leather
Vegan leather is usually made of PVC or polyurethane, which are synthetic fibres that have a similar environmental impact to polyester. It is certainly better for animal welfare, but it is not an eco-friendly option. However, some plant-based substitutes of leather exist, such as the pineapple fibre.

Eco-Textiles:

Eco-textiles refers to all fabrics, clothing and accessories that have been manufactured, produced, and processed in an environmentally-conscious manner. This manner reduces any negative impact in the form of pollution or damage to the planet.

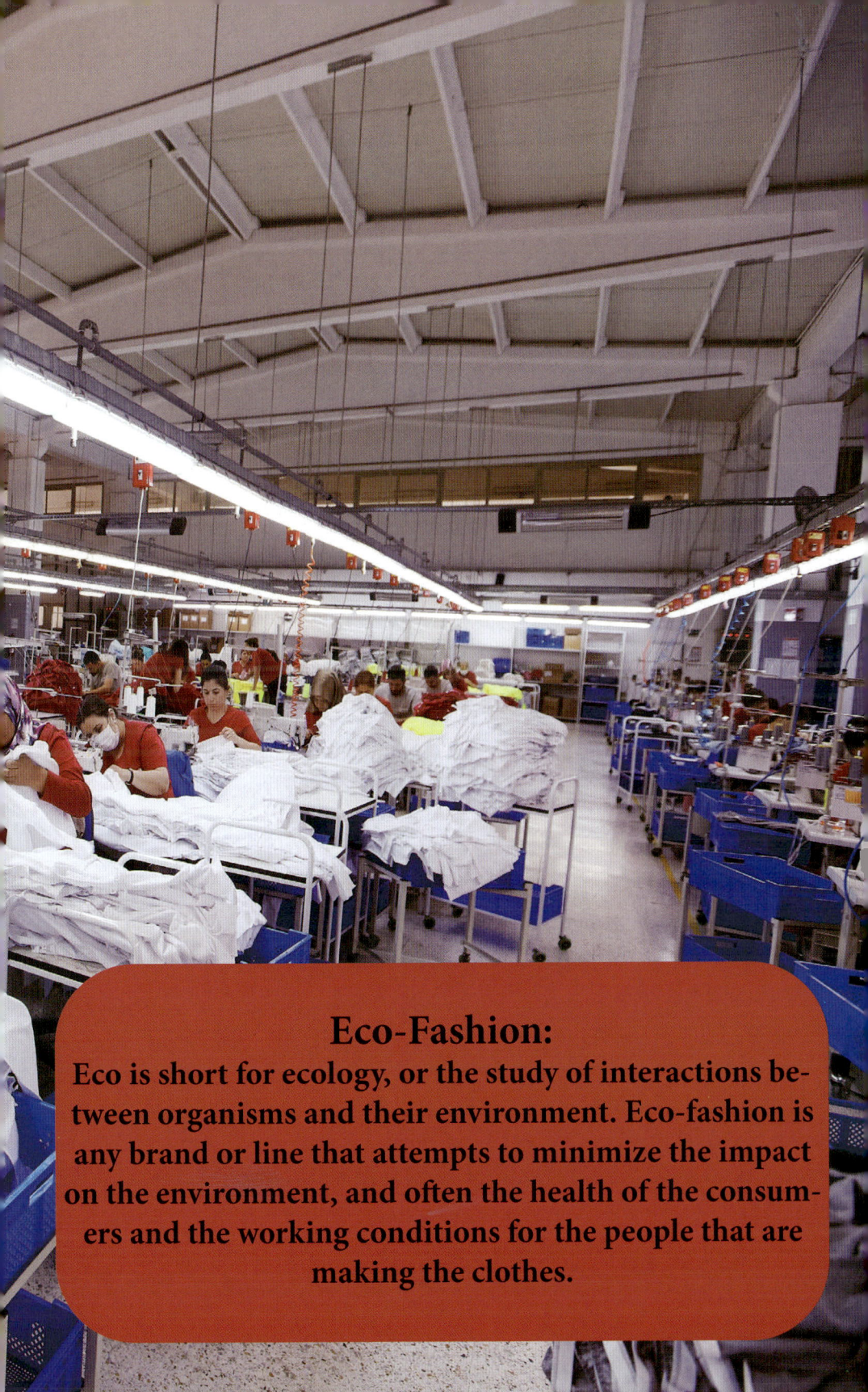

Eco-Fashion:
Eco is short for ecology, or the study of interactions between organisms and their environment. Eco-fashion is any brand or line that attempts to minimize the impact on the environment, and often the health of the consumers and the working conditions for the people that are making the clothes.

Step Four

Health & Beauty

INTERNATIONAL MANUAL OF CLIMATE CHANGE CONTROL (IMCCC) • 83

STEP FOUR

All about 'Eco-friendly' Health and Beauty Products

Health and beauty encompasses a variety of products, including fragrances, makeup, hair care and coloring products, sunscreen, toothpaste, and products for bathing, nail care, and shaving. The industry overlaps with other markets like chemical, health care, and petroleum.
In this book, Health and Beauty products divided into Three main catogries:

- Cosmetics
- Baby
- Hygine

Green Cosmetics:
The Push for Sustainable Beauty

As public interest in sustainability continues to climb, many cosmetic manufacturers are seeking more natural and environmentally-friendly emulsifiers and ingredients for their products. The benefits of "green" beauty products extend beyond trends — increasing studies show the toxicity of conventional cosmetics, and the natural cosmetics market continues to grow rapidly and consistently.

Manufacturing companies interested in venturing into the green market must know the details behind the sustainability movement, including the benefits of going green and the potential of the market. In modern marketing, the word "green" has become synonymous with "organic" or "healthy." When a consumer sees the phrase "green cosmetics," they will automatically make eco-friendly assumptions about the product or company.

But the field of green cosmetics still needs clarification. Typically, the term is used to describe products using environmentally-friendly formulations, production practices or packaging methods. In the United States, the Federal Trade Commission (FTC) has published guidelines to clarify what green or natural means in marketing terms, though these guidelines are still loosely defined.

With respect to the cosmetics industry, "green" and "sustainable" cosmetics are defined as cosmetic products using natural ingredients produced from renewable raw materials. Many companies use petrochemical ingredients derived from petrol, a non-renewable and economically volatile resource. Bio-based oleochemicals, on the other hand, derive from renewable plant and bacteria sources and are the crux of the green cosmetics movement.

How Are Sustainable Cosmetics Made?

Cosmetics developers worldwide are doggedly pursuing these oleochemicals, along with any potential sources for them. Some examples of common sources include:

Natural Oils: Palm and coconut oils are often used to derive fatty alcohols, which are used as chemical surfactants. Other oils include argan oil and avocado oil. Glycerine, a derivative of palm oil, is a common byproduct.

Agricultural Plants: Soybeans, corn and other agricultural plants are used throughout the cosmetic industry to produce oils and alcohols. Green cosmetic emulsifiers, surfactants and biocatalysts are derived using these plants, which can be cheaply and sustainably sourced.

Bacteria: One example of a renewable resource currently under development is the Deinococcus bacteria, a bacterium studied by Deinove in France for its chemical production properties. Deinove has used the bacterium to create aromatic ingredients and pigments for the cosmetic industry, representing a potential market value in the hundreds of millions of dollars. Manufacturers split these raw materials into oleochemicals at a processing plant. The fats or oils are divided by hydrolysis, which uses water, or alcoholysis, which uses alcohol.

Ingredients That Aren't Sustainable

Avoid many of the toxic elements found in popular brands. These chemicals damage environmental and human health, and consumers should never read them on a "green" label.

Aluminum:
Commonly used in antiperspirants, aluminum enters the body through the underarm tissue and blocks sweat ducts. However, it has also been linked to breast cancer, Alzheimer's disease and osteoporosis.

Dibutyl phthalate (DBP):
Often found in nail products, DBP is a solvent for dyes. Considered toxic to human reproduction, it enhances the ability of other chemicals to cause genetic mutations. While Canada has banned DBP from all children's toys, no action has yet been taken against its presence in cosmetics.

Coal tar dyes:
On labels, coal tar dyes are listed as p-phenylenediamine or colors titled "CI" and followed by a five-digit number. These dyes are mixtures of petrochemicals, and they have been linked to cancer in humans.

BHA and BHT:
BHA and BHT are synthetic antioxidants used as preservatives, and they are most common in lipsticks and moisturizing creams. The European Commission has released evidence that BHA and BHT disrupt the endocrine system.

Formaldehyde-releasing preservatives:
These preservatives are present in a wide range of cosmetics, as well as in cleaning products such as toilet bowl cleaners. As their name suggests, formaldehyde-releasing preservatives continuously release small amounts of formaldehyde, a known human carcinogen.

Examples of Sustainable Cosmetics:

Many manufacturers have found success using oleochemical-based products, and beyond creating high-quality and effective products, they have gained a loyal customer following. Here are some of the most well-known, sustainable cosmetics companies and their products:

Native: Native produces deodorants with organic, natural ingredients. Native has built their brand around "simple, nontoxic ingredients you can understand." Their oleochemical-derived ingredients include shea butter, coconut oil and castor bean oil.

Burt's Bees: From simple beeswax candles to a lip-product empire, Burt's Bees has become an international leader in sustainability. The company creates cosmetics and personal care products, and in addition to natural, organic ingredients, it has a "no-waste" manufacturing policy. They rely on botanical oils, herbs and beeswax to come up with their world-recognized products.

RMS Beauty: RMS Beauty provides a wide range of cosmetics, from foundation to mascara. Dedicated to using organic ingredients, RMS creates non-toxic makeup products that heal and protect the skin. They use low-heat processing to ensure their ingredients remain as natural as possible.

Blissoma: Focusing on skincare, Blissoma offers a large selection of products organized by skin type and need. Their preservative-free cosmetics include natural ingredients like fruit enzymes, Vitamin C and organic herbs and grains.

Drunk Elephant: Committed to using clean, natural ingredients, Drunk Elephant manufactures a range of sustainable skin care products. They have a devoted consumer following and strive to create products that are both clinically-effective and naturally-sourced.

It's possible for any company to incorporate green materials in their cosmetics. If you want to branch into the world of sustainable, oleochemical-derived products, begin with some of these safe and effective ingredients.

Fatty Acids: Fatty acids like coconut fatty acid, stearic acid and oleic acid are green ingredients used as lubricants, adhesives and release agents, as well as emulsifiers and base stock. You can incorporate naturally derived fatty acids into a wide range of cosmetic products, including soaps, ceramic powders, lotions and creams.

Castor Oil: Made by pressing the seeds of the castor plant, castor oil is a beneficial ingredient that has a range of anti-inflammatory and pain-relieving properties. When used in hair cosmetics, materials like Jamaican Black Castor Oil both remove impurities and clarify the scalp, resulting in more effective and more eco-friendly product.

MCT Coconut Oil: Extracted from the kernel of mature coconuts, MCT Coconut oil is a highly specialized and versatile carrier oil. Light, smooth and easily absorbed into the skin, MCT oil is especially useful in skincare products. Because it doesn't leave an oily residue, MCT oil is ideal for products marketed as oil-free or for sensitive skin types.

DMDM Hydantoin: A powerful antimicrobial agent, DMDM Hydantoin is a halogen-free preservative. This eco-friendly ingredient can be added to both rinse-off and leave-on products, including eye and skin creams, shampoo and conditioner, sunscreen, liquid soap and make-up remover.

Phenoxyethanol: Inhibiting both bacteria and mold growth, phenoxyethanol is an effective preservative used in a wide range of green cosmetics, from lotions and creams to make-up and gels. Phenoxyethanol serves a variety of roles within cosmetics, including solvent, fixative and topical anesthetic functions.

Committed to sustainable manufacturing, Acme-Hardesty offers each of these green ingredients for manufacturers in the cosmetics industry.

Why Buy Natural and Sustainable Cosmetics

Environmental Responsibility

Modern consumers have a growing global consciousness, and they care about social and environmental responsibility. One of the main benefits of sustainable products is their kinder environmental impact.

Every week, new stories surface about dangerous carbon outputs or vast plastic floats in the ocean. Many petrochemicals in conventional cosmetics are toxic pollutants and degrade the environment as well as our bodies. As we become more ecologically aware, consumers demand natural, low-polluting products.

A recent example of pollution and consumer demand is the ban of microbeads. Microbeads are tiny pieces of plastic found in many shower scrubs and exfoliating products. However, they do not dissolve, and in 2015, a study reported that over eight trillion microbeads were being washed into our waterways every day. Later that year, U.S. President Barack Obama signed a bill banning the small plastics, illustrating that environmental stewardship is an increasing priority to the nation and its consumers.

Increased Effectiveness

Natural and oleochemical ingredients are less likely to cause skin irritation or allergic reactions. Without synthetic, toxic chemicals or artificial colors, sustainable products rely on the healing properties found naturally in plants and animals — the ingredients humans have been using for centuries. Consider glycerine, a natural derivative of palm oil. The clear, non-toxic liquid is used in soaps, pharmaceuticals and cosmetics. Since it is a humectant, glycerine can retain water, making it an excellent moisturizer. Glycerine enhances the body's hygroscopic characteristics, encouraging the skin to absorb and hold on to water. As a non-irritating substance, it can be applied anywhere on the body. It is an effective anti-aging ingredient and, due to its anti-microbial properties, can also serve as an acne treatment.

Long-Term Health

While petrochemicals may deliver short-term results, the long-term effects can be highly toxic to humans and the environment. Years of synthetic cosmetics use has been traced to headaches, eye damage, acne, hormonal imbalance and premature aging. Phthalates have even been linked to cancer and type II diabetes. By choosing sustainable cosmetics, a consumer forgoes the stress and uncertainty of toxic, synthetic products and invests in their long-term health and beauty.

Why Produce Green Cosmetics?

1. Improved Product Quality
High-quality cosmetics provide effective results without putting the consumer at risk. However, many petrochemical products, like mineral oil, present a low level of toxicity to users. When aerosolized and inhaled, such products have been shown to be allergens and, as some studies suggest, may cause cancer. With most bio-based products, the toxicity to the end-user is reduced, creating safer, higher-quality products.

2. Enhances Brand Reputation
Green products send a message to consumers — this company is committed to quality, safety and sustainability, and is worthy of your trust. As more and more people grow concerned about synthetic products, consumers are looking for companies that practice transparency and honesty. By moving towards sustainable, green products, you show your global and social awareness. This promotes customer loyalty to a brand, not just to products. People will begin — and continue — to purchase a company's products because they agree with its mission.

3. Increases Corporate Responsibility
Green cosmetics also present a unique opportunity for cosmetics manufacturers to focus on corporate responsibility. In addition to the positive impacts green marketing can have on a company's image, taking the extra steps of sustainable sourcing or packaging can also make a significant impact. When a company increases its sustainability initiatives, it takes ownership for its impact on global health and economies. By taking corporate responsibility for its manufacturing, a business gains authority and respect among consumers as well as suppliers and other members of the distribution chain.

The Future of Sustainable Cosmetics

Manufacturers shifting to sustainable cosmetics production have a promising future. The growing interest in sustainable cosmetics has had a significant effect on the cosmetics market. With an increasing number of consumers and retailers demanding cosmetics with natural or sustainable ingredients, the green cosmetics market has experienced a 15 percent annual growth rate.

This growth rate far outpaces that global personal care and cosmetics industry, which is currently sustaining an overall 5 percent annual growth rate. By 2025, the organic beauty market will reach $25.11 billion. Within the personal care industry, the oleochemicals market is increasing as cosmetic manufacturers continue to turn away from petrochemicals. Fatty acids, in particular, should experience boosts on the green side of the market, considering that they accounted for 57 percent of the total oleochemical product demand in 2013.

The Asia-Pacific region is an area of particular interest for this market since the region accounted for 41.9 percent of the total oleochemicals market in 2013 for its abundance of raw materials and large consumer base. Both figures are unsurprising considering the quantities of bulk cosmetic glycerine regularly exported from the region. As petrochemicals continue to experience volatility in the market, turning to sustainable material sources may be the best long-term decision for cosmetics manufacturers worldwide.

Consumers are increasingly demanding sustainable products that are not toxic to themselves or the environment. The natural market is growing exponentially, and choosing raw, natural materials will cement your brand as a safe choice — both environmentally and economically.

EU and UK Cosmetic product definition:
Based on the definition of the cosmetic products, products that may seem to be cosmetics, like nail wraps, a comb or a toothbrush, therefore aren't cosmetics, even though they are placed in contact with the external parts of the human body, and their primary function is to change appearance, but they wouldn't be considered a substance or a mixture.

> **A cosmetic product in Europe and UK is defined in the Regulation 1223/2009 as follows:**
> 'cosmetic product' means any substance or mixture intended to be placed in contact with the external parts of the human body (epidermis, hair system, nails, lips and external genital organs) or with the teeth and the mucous membranes of the oral cavity with a view exclusively or mainly to cleaning them, perfuming them, changing their appearance, protecting them, keeping them in good condition or correcting body odours. (EU Regulation 1223/2009, Article 2.1.a)

Since products have to be placed in contact with the external parts of the human body or with the teeth and the mucous membranes of the oral cavity, any product intended to be ingested, inhaled, injected or implanted into the human body would also not be considered a cosmetic product in the EU or the UK. Breast implants then aren't cosmetics, even though their primary function is also to change appearance.

CLASSES OF COSMETIC PRODUCTS:

Cosmetic product may include:

creams, emulsions, lotions, gels and oils for the skin,
face masks,
tinted bases (liquids, pastes, powders),
make-up powders,
after-bath powders,
hygienic powders,
toilet soaps,
deodorant soaps,
perfumes, toilet waters and eau de Cologne,
bath and shower preparations (salts, foams, oils, gels),
depilatories,
deodorants and antiperspirants,
hair colorants,
products for waving, straightening and fixing hair,
hair-setting products,
hair-cleansing products (lotions, powders, shampoos),
hair-conditioning products (lotions, creams, oils),
hairdressing products (lotions, lacquers, brilliantines),
shaving products (creams, foams, lotions),
make-up and products removing make-up,
products intended for application to the lips,
products for care of the teeth and the mouth,
products for nail care and make-up,
products for external intimate hygiene,
sunbathing products,
products for tanning without sun,
skin-whitening products,
anti-wrinkle products

The Cosmetic Product has to be intended to place in contact with:

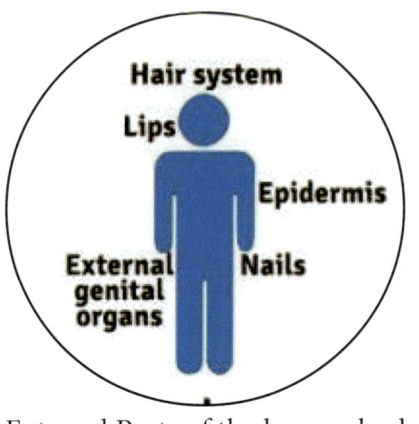

External Parts of the human body

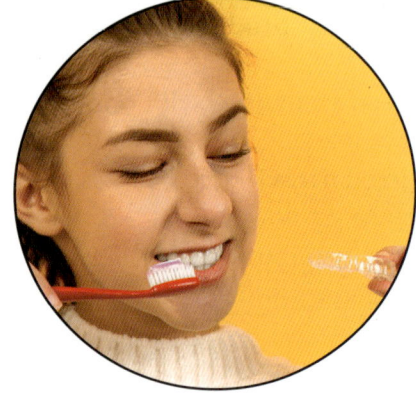

Teeth

Mucous membranes of the oral cavity

The assessment of whether a product is a **ECO friendly** cosmetic product has to be made on the basis of a case-by-case assessment, taking into account all characteristics of the products.

INTERNATIONAL MANUAL OF CLIMATE CHANGE CONTROL (IMCCC)

Cosmetics FAQs:
What is meant by vegan cosmetics?
When a beauty product is marked as vegan, it means the same thing a vegan diet does. This label infers that the product does not contain any animal or its byproducts in it.

What is the difference between vegan makeup and regular makeup?
Most often, the difference between vegan and vegetarian cosmetics is that products labeled "vegetarian" will contain natural animal-made ingredients, most commonly beeswax and honey, which are avoided by vegans. Many green beauty brands that aren't 100% vegan will state that they're vegetarian. ... Testing Ingredients.

What is cruelty free in cosmetics?
Cruelty-free cosmetics is a category containing all cosmetics that have not been tested on animals. ... Many companies brand themselves as cruelty-free but still use raw materials that have been tested on animals.

"Cruelty-free" can be used to imply that:

- Neither the product nor its ingredients have ever been tested on animals. This is highly unlikely however, as almost all ingredients in use today have been tested on animals somewhere, at some time, by someone — and could be tested again.
- While the ingredients have been tested on animals, the final product has not.
- The manufacturer itself did not conduct animal tests but instead relied on a supplier to test for them — or relied on another company's previous animal-test results.
- Either the ingredients or the product have not been tested on animals within the last five, ten, or twenty years (but perhaps were before, and could be again).
- As in the case of the CCIC's Leaping Bunny Program, neither the ingredients nor the products have been tested on animals after a certification date and will not be tested on animals in the future.

Can skin care products cause allergic reaction?
Personal care products like makeup, skin cream, and fragrances also commonly cause rashes. It's not well understood how chemical compounds in personal care products trigger these rashes, called allergic contact dermatitis.

Can you suddenly become allergic to something?
Allergies can develop at any point in a person's life. Usually, allergies first appear early in life and become a lifelong issue. However, allergies can start unexpectedly as an adult. A family history of allergies puts you at a higher risk of developing allergies some time in your life.

How to Prevent Cosmetic Allergies?
Avoid ingredients in cosmetics and skincare products that are irritating to your skin. Look for labels such as "hypoallergenic", "sensitivity tested", "paraben-free", "phthalate-free", "non-comedogenic" and "fragrance-free".

What is a Non-GMO?
Non-GMO means a product was produced without genetic engineering and its ingredients are not derived from GMOs.

What is Paraben free?
Parabens can act like the hormone estrogen in the body and disrupt the normal function of hormone systems affecting male and female reproductive system functioning, reproductive development, fertility and birth outcomes. While nearly all beauty products use some kind of preservatives to make their products last longer, paraben-free cosmetics may be safer to use. The term "paraben-free" is meant to let consumers know that these harmful chemicals aren't a part of the product formula.

What is SLS ?
Sodium Lauryl Sulfate (SLS) strips the skin of its natural oils which causes dry skin, irritation and allergic reactions. It can also be very irritating to the eyes. Inflammatory skin reactions include itchy skin and scalp, eczema and dermatitis.

The Most Common Allergens in Health & Beauty Products;

Fragrance
Fragrance is one of the most common causes of allergic contact dermatitis in skincare products. "Fragrance can contain hundreds of components, and companies are not required to disclose all the ingredients that make up what's labeled as 'fragrance,'" he says. Fragrance-free is also not the same as unscented (unscented products might contain masking fragrances that neutralize perfume). Between the two, "fragrance-free is the better option if you have sensitive skin."

Parabens
"Parabens are a group of synthetic compounds commonly used as preservatives in a wide range of personal care products," . "They might cause an allergic reaction in certain people, and are more likely to irritate those with existing skin issues like eczema, psoriasis, and contact dermatitis."

Sulfates
Sodium laureth sulfate and sodium lauryl sulfate are two common skincare, bath, and hair product ingredients that also may cause rashes and itching.

Dyes
Dyes—most often found in hair products and more pigmented cosmetics—are another culprit; The dye ingredient which most often causes allergic reactions is paraphenylenediamine (PPD).

Benzyl Alcohol
"Benzyl alcohol is used for its fragrance, preservative abilities, and antimicrobial action, "In rare cases, it can cause a hive-like reaction."

Propylene Glycol
Propylene glycol is often used in moisturizers as a humectant to lock in moisture—and even at low concentrations, allergic reactions can occur, he warns.

Essential Oils

Last but not least? Essential oils. "They're highly concentrated substances that are extracted from various trees and plants for their fragrance and antimicrobial action," Tea tree oil is the most common essential oil allergen. Steer clear of products containing essential oils if your skin tends to be sensitive.

Today's sustainable paper eco friendly cosmetic packaging offers a perfect material for products that can be reduced and reused, are 100% recyclable and fully biodegradable.

Step Five

Cleaning

STEP FIVE

All about 'Eco-friendly' Cleaning Products

Green cleaning products should not contain hazardous chemicals, and so they are likely to pose fewer health risks. Green cleaning products are less hazardous for the environment, too. They do not contain chemicals that cause significant air or water pollution and are often in recyclable or recycled packaging. The eco-friendly cleaning products, also known as green cleaning products, are made from plant-based ingredients, natural colors or fragrances, uses eco-friendly packaging methods, and are biodegradable. Support sustainable human and ecological use and reuse of remediated land; Minimize impacts to water quality and water resources; Reduce air toxics emissions and greenhouse gas production; Minimize material use and waste production; and Conserve natural resources and energy.

How do I know which cleaning products are the most environmentally friendly?

Almost all people from all over the world use household cleaning products from dish detergents to bathroom cleaners and floor polish to scouring pads. Most of us are exposed to cleaners on a daily basis,... Even if we don't use cleaners, it's likely we're regularly come into contact with them at work, school or elsewhere.

Unfortunately, cleaners often contain harsh chemicals that can be harmful to our health and planet. Health effects associated with cleaning products include asthma, contact dermatitis, burns to the skin and eyes and inflammation or fluid in the lungs. Long-term repercussions may include reproductive problems, cancer, heart disease and other health issues. The environment also can fall victim to cleaning products' acrid ingredients. Chemicals in laundry detergents, for example, have been found in 75 percent of streams

and waterways throughout in different countries. Some ingredients in cleaners have been directly linked to environmental problems, such as chemicals getting into bodies of water and foaming in streams, and some commonly used household cleaner ingredients have room for improvement even today. Health and environmental concerns have prompted many consumers to push for safer alternatives to cleaning products. **But identifying environmentally safe cleaners can be challenging for consumers.** With so many product options, choosing the safest, healthiest cleaners for the home can be challenging for reasons other than too many choices, namely the lack of a national regulatory body. A fraction of the tens of thousands of chemicals in commerce in different countries are used in consumer goods like household cleaners. Chemicals are regulated as they enter commerce rather than at the product level.

But shoppers who want to know exact ingredients might not find what they're looking for on household cleaner labels. In many countries law does not require manufacturers of cleaning products to list all ingredients on labels. But manufacturers might be changing their ways in the near future. Some of the associations, launched a joint, voluntary effort to encourage their members to list their ingredients in a public format. In the meantime, consumers are left to make sense of what's on the packaging.

Different manufacturers can make the same marketing claim like "degradable" or "ozone-friendly" and mean different things with those terms. This has resulted in confusion among consumers.

Besides looking for the Design for Environment Labellings and knowing what marketing terms mean, consumers can also read product packaging to make sure environmental claims are qualified. All assertions should specify whether it's referring to the product, packaging or both. Similarly, Environmental labels should come with an explanation and identify the third party doing the certifying.

The organization should be independent from advertisers and have expertise in the area for which it's certifying. Other indicators of environmental responsibility are the following: recycled, recyclable or refillable containers; concentrated products that require less packaging; cleaners free of chlorofluorocarbons (CFCs) that can deplete the ozone; and degradable, biodegradable or photodegradable product contents or packaging.

For example, a toilet cleaner ad that claims the solid waste generated by disposing of its container is "now 20 percent less than our previous container," is in good practice if the cleaner company can prove disposal of the new package contributes 20 percent less waste by weight or volume to the solid waste stream. Comparatively, the general claim "20 percent less waste" is ambiguous and therefore deceptive because it's unclear if the claim is referring to a preceding product or that of a competitor, according to the general principles on International Environmental labelling.

Homemade Cleaning Products: Natural, Green, Eco-Friendly

A mixture of vinegar and baking soda can do wonders for your cleaning needs. This combination can be used in many ways to fight against severe stains, so you do not need to run out to the grocery store to buy a solution filled with chemicals anymore. Not only will natural cleaners make your life better, they will virtually eliminate that bad smell in the house and they're surprisingly inexpensive to create.

Chemical - free Recipes for Homemade Cleaning Products:

If you're wanting to pitch those toxic, commercial household cleaners and switch to natural, homemade cleaner, these simple recipes will have you cleaning green in no time, Before we get to the cleaning, let's check out some of the most common (and most useful) non-toxic cleaning products:

Baking Soda
Baking soda is a pantry staple with proven virus-killing abilities that also effectively cleans, deodorizes, brightens, and cuts through grease and grimeTrusted Source.

Castile Soap
Castile soap is a style of soap that's made from 100 percent plant oils (meaning it uses no animal products or chemical detergents).

Vinegar
Thanks to its acidity, vinegar is nothing short of a cleaning wunderkind—it effectively (and gently!) eliminates grease, soap scum, and grime.

Lemon Juice
Natural lemon juice annihilates mildew and mold, cuts through grease, and shines hard surfaces (It also smells awesome.).

Olive Oil
This good-for-you cooking oil also works as a cleaner and polisher.

Essential Oils
Essential oils have gained popularity thanks to aromatherapy, but these naturally occurring plant compounds also make great scent additions to homemade cleaning products (particularly if you're not into the smell of vinegar). Essential oils are generally considered safe, but these extracts can trigger allergies—so keep this in mind when choosing scents.

Borax

Many DIY cleaners tout Borax (a boron mineral and salt) as a non-toxic alternative to mainstream cleaning products; however, the issue is pretty hotly debated. Some research suggests Borax can act as a skin and eye irritant and that it disrupts hormones. For this list, we've chosen to avoid products that use Borax.

A note on mixing products:

Most of these ingredients can be used in combination with each other; however, many sources advise against mixing castile soap with vinegar or lemon juice. Since castile soap is basic (i.e., high on the pH scale) and vinegar and lemons are acidic, the products basically cancel each other out when used in combination (though it's fine to wash with a base—like castille soap—and rinse with an acid—like vinegar!).

Cleaning Recipes

Many of these cleaners can be used in multiple places, but we've assigned them to particular areas for easy reference:

- **Bathroom**
- **Kitchen**
- **Lundry Room**
- **Others**

Bathroom

1. Toilets

For a heavy-duty toilet scrub that deodorizes while it cleans, pour ½ cup of baking soda and about 10 drops of tea tree essential oil into the toilet. Add ¼ cup of vinegar to the bowl and scrub away while the mixture fizzes.

For daily cleaning, fill a small spray bottle with vinegar (about 1 cup should do it) and a few drops of an essential oil of your choosing (lemon and tea tree both work well). Spray on the toilet seats, let it sit for a few minutes, and then wipe the surface clean.

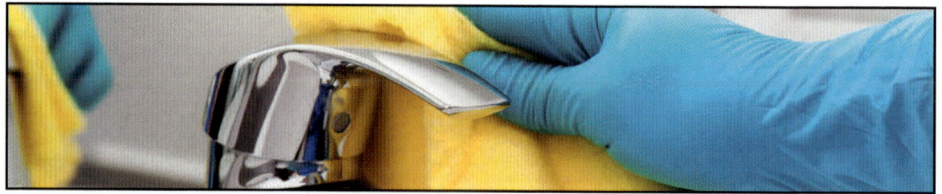

2. Tub and Shower

Tubs and showers can produce some of the toughest grime, but it's no match for the cleaning power of vinegar. To get rid of mildew, spray pure white vinegar on the offending area, let it sit for at least 30 minutes, and then rinse with warm water (don't be afraid to use a sponge if rinsing doesn't clear away the grossness on its own). Alternatively, try mixing together baking soda with a bit of liquid castile soap, then scrub and rinse.

For daily cleaning or to get rid of soap scum, mix 1 part water with 1 part vinegar (and a few drops of essential oils if you're not into the smell of vinegar) in a spray bottle. Spray, let it sit for at least several minutes, and then wipe away.

3. Disinfectant

Skip the bleach and make a homemade germ-killer instead. Just mix 2 cups of water, 3 tablespoons of liquid soap, and 20-30 drops of tea tree oil. Voila!

4. Air Freshener

Defeat less-visible bathroom "uncleanliness" with this homemade, non-toxic air freshener. All you need is baking soda, your favorite essential oil, and an old jar with a lid you don't mind poking holes in (follow the link for full instructions).

5. Hand Soap

Once you're done cleaning the bathroom, it's time to make yourself clean (or at least your hands). To make a non-toxic, foaming hand soap, mix together liquid castile soap and water (and an essential oil if you feel like it) in a foaming soap dispenser. Fill about one fifth of the bottle with soap, then top it off with water.

Kitchen

6-Countertops

For a simple, all-purpose counter cleaner, mix together equal parts vinegar and water in a spray bottle. If your countertop is made from marble, granite, or stone, skip the vinegar (its acidity is no good for these surfaces) and use rubbing alcohol or the wondrous power of vodka instead.

7. Cutting Boards

Talk about non-toxic: All that's needed to clean and sanitize cutting boards (wood or plastic) is… a lemon! Cut it in half, run it over the surfaces, let sit for ten minutes, and then rinse away. If you need some serious scrubbing power, sprinkle some coarse or Kosher salt over the board, and then rub with ½ a lemon.

8. Oven

To clean stubborn, caked-on food out of the oven, just heat the over to 125 degrees and grab your spray bottle of vinegar (see "countertops" above). Once the oven is warm, spray the caked-on stuff until it's lightly damp and then pour salt directly onto the affected areas. Turn off the oven, let it cool, and then use a wet towel to scrub away at the mess. If that doesn't cut it, follow the same instructions but try use baking soda in place of salt (just let it sit for a few minutes before scrubbing).

9. Garbage Disposal

This one is so cool. Pour 1 cup of vinegar into an ice cube tray and top off the slots with water. Once they're frozen, toss a few down the disposal and let it run—doing so should remove any food that was stuck to the blades.

10. Microwave

It's easy to overlook the microwave while cleaning, but man can it get gross in there. To combat the gunk, pour some vinegar into a small cup and mix in a little lemon juice (exact amounts don't really matter). Put the cup in the microwave, let the microwave run for 2 minutes, and leave the door closed for several more minutes. Finally, open the door and simply wipe down all the sides with a warm cloth or sponge—no scrubbing required!

11. Sink Drain

To unclog a stuffed-up drain, start by boiling about 2 cups of water. Pour ½ cup of baking soda into the drain, and then add the water while it's still nice and hot. If that doesn't do the trick, follow the baking soda with ½ cup of vinegar, cover it up tightly (a pot lid should work nicely), wait until the fizzing slows down (when baking soda and vinegar come in contact, they'll react by fizzing) and then add one gallon of boiling water.

12. Pan De-Greaser

To cut through the grime on frying pans, simply apply some salt (no water necessary) and scrub vigorously.

13. Cast-Iron Pans

Kitchen professionals are pretty against using soap, steel wool, or dishwashers to clean cast-iron pans. Luckily, there's an alternative way to tackle cast-iron grossness: combine olive oil and a teaspoon of coarse salt in the pan. Scrub with a stiff brush, rinse with hot water, and you're done!

14. Dishwasher Detergent

If you're lucky enough to have a dishwasher, simply mix together 1 cup of liquid castile soap and 1 cup of water (2 teaspoons of lemon juice optional) in a quart-size glass jar. Add some of this mixture to one detergent compartment of the dishwasher, and fill the other compartment with white vinegar.

15. Dish Soap

If washing dishes by hand, simply combine 1 cup of liquid castile soap and 3 tablespoons water (a few drops of essential oil optional) in a bottle of your choice. Shake well and use like you would any other dish soap.

16. Refrigerator Cleaner

To clean what is perhaps the toughest of all kitchen "gross spots," reach for the baking soda. Add about ½ cup of the white stuff to a bucket of hot water. Dip a clean rag in the mixture and use it to wipe down the fridge's insides.

17. Bleach

For serious disinfectant power, mix ½ cup baking soda, 1 teaspoon castile soap, and ½ teaspoon hydrogen peroxide. Use a cloth to apply the mixture to a wet surface, scrub, and then rinse thoroughly.

Lundry Room

18. Laundry Detergent
It's tough to come by homemade laundry detergents that don't use Borax, but give this one a try. The recipe calls for glycerin soap, washing soda, baking soda, citric acid, and coarse salt. For full instructions, follow the link!

19. Fabric Softener
Skip the liquid fabric softener and make clothes nice and snuggly the non-toxic way. Make a big batch of softener by adding 20-30 drops of the essential oil of your choice to a one-gallon jug of white vinegar. Add 1/3 cup to each laundry load (just be sure to shake the mixture prior to each use).

20. Laundry "Scenter"
To add a fresh, clean scent to laundry, make a sachet stuffed with your favorite dried herbs (lavender, peppermint, and lemon verbena are all great options). Toss it in the dryer while it's in use, and voila: customized, non-toxic scent!

21. Bleach
For a nontoxic laundry bleach alternative, add some lemon juice to the rinse cycle.

Everything Else

22. Floors
For a simple, effective tile floor cleaner, simply combine one part white vinegar with two parts warm water in a bucket. Use a mop or rag to scrub down the floors with the solution. No need to rinse off! (Note: this one's not recommended for wood floors).

23. Walls
To scrub down walls, mix ¼ cup white vinegar with 1 quart warm water, then use a rag to scrub those walls down. To remove black marks, simply scrub at the spot with a little bit of baking soda.

24. Windows and Mirrors
For an all-purpose window cleaner, combine 1 part white vinegar with 4 parts water (feel free to add some lemon juice if you're feeling citrusy), then use a sponge or rag to scrub away.

25. Furniture Polish
For an all-purpose furniture polish, combine ¼ cup vinegar with ¾ cup olive oil and use a soft cloth to distribute the mixture over furniture. For wood furniture (or as an alternative to the first recipe), combine ¼ cup lemon juice with ½ cup olive oil, then follow the same procedure.

26. Silver Cleaner

Put silver utensils and jewelry back to good use the non-toxic way. Line a sink or bucket with aluminum foil, lay out the silver on top of the aluminum, and pour in boiling water, 1 cup of baking soda, and a pinch of salt. Let it sit for several minutes and watch as—like magic—the tarnish disappears! Note: If you're concerned about immersing a particular item, simply rub it with toothpaste and a soft cloth, rinse it with warm water, and allow it air to dry.

27. Wood Cleaner

Clean varnished wood by combining 2 tablespoons of olive oil, 1 tablespoon of white vinegar, and a quart of warm water in a spray bottle. Spray onto wood and then dry with a soft cloth. (Note: Since olive oil can leave behind some (slippery) residue, this one might not be the best option for wood floors.)

Important Note: We've done our absolute best to provide the best information possible, but since we haven't tried every single one of these solutions in every possible cleaning situation, we can't vouch for them 100 percent.

What's the difference between products that disinfect, sanitize, and clean surfaces?

Products used to kill viruses and bacteria on surfaces are considered as antimicrobial pesticides. Sanitizers and disinfectants are two types of antimicrobial pesticides.

Cleaning	Cleaning removes dirt and organic matter from surfaces using soap or detergents.
Sanitizing	Sanitizing kills bacteria on surfaces using chemicals. It is not intended to kill viruses.
Disinfecting	Disinfecting kills viruses and bacteria on surfaces using chemicals.

Experts agree that frequent handwashing is one of the first lines of defense against many illnesses. But no matter how many times you wash your hands, there are always some sneaky little germs lurking around to hitch a ride on your skin. They loiter on shopping cart handles, linger on light switches, lurk about the phone and even hang around on the remote controls. That's why disinfectants and disinfecting cleaners can be a helpful option.

Why Disinfect
- Regular cleaning products do a good job of removing soil and many germs. Disinfectants or disinfectant cleaners are able to go further and kill many of those germs.
- Surfaces may be contaminated even when they're not visibly soiled.
- Germs can live on surfaces for hours or even days.

How to Disinfect

- Read the label before using any cleaning or disinfecting product to ensure you are following the directions for use and storage instructions.
- Pre-clean any surfaces prior to disinfecting to remove any excess dirt or grime.
- Apply the disinfectant, then the surface needs to stay wet for the entire time indicated on the product label; this is called contact time.
- If disinfecting food contact surfaces or toys, rinse with water after they air dry.
- When disinfecting, target surfaces that are frequently touched, especially if someone in the home is ill.
- If using a disinfectant wipe, throw out after using. Do not flush any non-flushable products.

What to Look for in a Disinfectant

Products that say "Disinfectant" on the label are required to meet legal specifications. To be sure the product has met all defined requirements for effectiveness, look for Registration Number on the label. You must follow the product label instructions exactly for the disinfectant to be effective. Your choices include:

Chlorine bleach. It disinfects when mixed and used properly. Read the label for instructions.
Disinfectant cleaners. These dual purpose products contain ingredients that help remove soil as well as kill germs.
Disinfectants. These products are designed to be effective against the germs indicated on their labels. Surfaces should be clean prior to disinfecting.

Some of the more frequently used active ingredients are sodium hypochlorite, ethanol, pine oil, hydrogen peroxide, citric acid and quats (quaternary ammonium compounds).

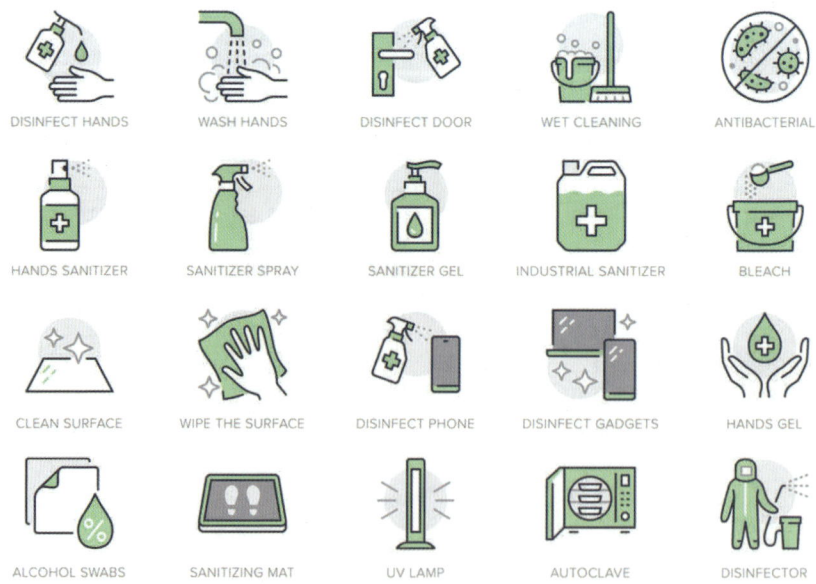

When killing surface germs is your goal, look for products that are disinfectants, some common disinfectants include Quaternary Ammonium Compounds (QACs), commonly referred to as Quats.

The Future of Cleaning

Cleaning technology has come a long way from the ancient Babylonian way of soap-making. Today's cleaning products are the result of thoughtful design, experimentation, and safety testing.

The machines we use to clean have also improved, becoming more sustainable and friendly for our environment. So far we have been able to make new cleaning products that allow us to wash in cold water (saving energy from water heating), wash with less water, and make packaging smaller (to save material and avoid shipping extra weight).

Future scientists will have a great opportunity to continue to create new cleaning design products that will continue to keep us healthy and do even more to help protect human health and the environment.

Nanotechnology is the future, not only for things like computers and building materials but also for cleaning products. Not only is it important to focus on keeping surface areas clean, but we should also keep in mind air quality. Many workers have transitioned to more remote work over the past year, meaning more time spent inside your home with who knows what kind of germs or contaminants lingering. Some air-purification systems don't always catch pollutants at room temperature.

Nanotechnology aids in the filtration of such pollutants. Air-purifier sprays like Purbloc's NANO GRAB are your ticket to a clean, germ-free, home.

Harmful toxins leached out by ordinary household cleaners can cause everything from chapped hands to potential lung issues in the long term. Long-term health issues can then translate to high costs.

By going with a natural cleaner you're thinking of not only your long-term health, but you're saving money in the long run by taking care of yourself now and those around you with safe and effective supplies.

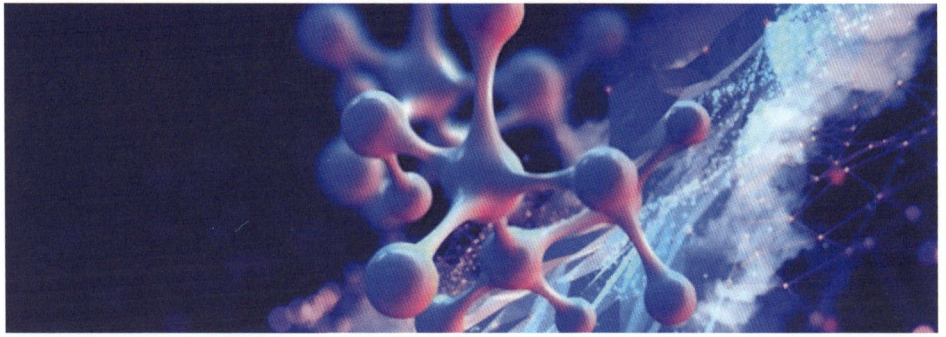

Keep in Mind
There are so many incredible real-world benefits of nanotechnology. A summation of a few products and processes enhanced by such benefits:

- Coatings to reduce cleaning efforts
- Improved energy efficiency
- Fabrics with increased resistance to stains
- Extra durable goods that increase the life of a product thus creating less waste in the long run
- Easier and more efficient water filtration
- Insulation materials reducing the energy needed to heat and cool buildings

Families and businesses alike can all benefit from the efficient uses mentioned above. From the reduction of energy consumption to the wonders of regenerative medicine, nanotechnology is a key player in the race toward enhancing and caring for our way of living.

What makes cleaning products eco-friendly?
Not containing toxic chemicals.
Has biodegradable ingredients.
Being made from plants.
Not using single use plastic.
Being plastic free entirely.
Being refillable.
Not creating waste.
Being vegan and/or cruelty free.

ECO HO
nature frie

SEHOLD
y products

Step Six

Stationery

STEP SIX

All about 'Eco-friendly' Wood & Stationery Products

Eco-friendly stationery is a broad category that encompasses recycled and sustainable stationery, along with zero-waste gifts like eco-friendly notebooks that can be fully recycled once they have been used. The term applies to all kinds of green stationery, from eco-friendly pens and pencils, to recycled notebooks and ethical stationery gifts – and different products might have different ethical and environmental credentials. For example, some eco stationery may be made from ethically produced materials like sustainably managed timber, while others might use more recycled raw materials or be designed in such a way to avoid ending up in a landfill site.In this volume, we'll look in more detail at what makes green stationery 'green' and why it makes sense to buy eco-friendly stationery for yourself and others.

What is sustainable stationery?
There's no strict definition of sustainable stationery. It could be recycled or recyclable (or both), it might use raw materials that would otherwise go to landfill, or it might be ethical in some other way. Because it's such a broad definition, there's plenty of choice, allowing you to choose a sustainable pen that fits your lifestyle and your own eco priorities.

Why should I try sustainable stationery?
There's really nothing to lose. Sustainable stationery often does not cost any more than its traditional equivalents, and there are some really stunning designs of eco pens, pencils and notebooks to choose from.

YOU CAN MAKE A CONSCIOUS EFFORT TO BE ENVIRONMENTALLY FRIENDLY

If you want to take action for sustainable living, without making major changes to your lifestyle, eco stationery is a simple first step to take. Sustainable stationery products function exactly as normal and there's usually no visible difference either, so you get peace of mind without compromise.

IT SETS A GOOD EXAMPLE
As well as improving your personal eco-profile, you can encourage others to take their own action for sustainable living too. International Environmental Labelling Book series of 11 volumes is a great gift for this scenario; by giving it to friends, family and co-workers. You can give this book series (look at the end of this book for more detail), an eco-friendly notebook, pen or pencil and send them on their own path towards a more eco-friendly life-style.

YOU WILL REDUCE WASTE
Sustainable stationery is typically designed to last longer. In fact, any refillable pen will generate less waste than a disposable pen, so opt for a cartridge pen, traditional fountain pen or refillable ball pen if you want to do your bit.

> If you are in business, you could consider giving a customised eco-friendly gift, to encourage your customers to cut down on their use of disposable stationery too.

How does eco-friendly stationery help the environment?

Eco-friendly stationery (depending on the product) can divert materials away from landfill during its manufacture, and reduce landfill waste due to disposable plastic pens being thrown away. Sustainable stationery is also more likely to use renewable raw materials such as wood and eco plastics, rather than single-use plastics made from fossil fuels.

How is sustainable stationery made?

Green stationery is made from sustainable raw materials – and these, in turn, are produced using ethical, long-term sustainable practices. For example, One of the the largest pencil brand in the world, has a pioneering forestation project in Brazil and Colombia where two million trees are planted every year. Whenever a tree is felled to provide timber for more than the 2.2 billion pencils manufactured each year, a new one is planted. This is just one example of how sustainable stationery can be made and managed.

Which sustainable stationery products should I try?
Ecolabel programs authorize the use of environmental logos on products or services that meet a strict set of criteria. These ecolabels indicate an overall environmental preferability of a product or service within a particular product or service category based on life cycle "considerations," although not necessarily a more complex full life cycle assessment. Some ecolabels are created and managed on a national level while others are international in scope. They may be administered by government bodies or private sector labelling standards organizations, and typically involve certification by legitimate and independent third party organizations.

Final thoughts
We could all live a little more sustainably, but simple steps like using more sustainable stationery allow us to do so without compromising on our existing lifestyles in any significant way. Many people have already swapped plastic drinking straws for metal, bamboo or paper equivalents. Switching to green stationery is an obvious next step. And with such a great selection of eco pens and pencils, eco-friendly notebooks and zero-waste gifts available, there's no reason to delay investing in a great-quality pen or pencil that will stay by your side for many years to come.

19 Eco Friendly And Zero Waste School Supplies

New stationery and school supplies has always been an exciting time for most kids (and many adults too!). Writing this article brings back memories of fresh stationery and the joys of organizing and reorganizing pens, pencils, rulers, notebooks…Although, for parents and adults who are aiming to keep school sustainable though, it can bring a sense of dread. Plastic packaging, toxic components, and boxes of unused junk. To help, here are a few strategies for sustainable, low or zero waste school supplies and stationery:

Use what you already have. Scrounge up loose pens and forgotten notebooks (they're usually full of mostly blank pages!).
For things you know you no longer need, consider donating any school supplies that still have a useful life and recycling those that don't.
See what you can find secondhand. eBay is a good place to look for secondhand online school supplies like books, graphing calculators, and other technology (now that many classrooms and home learning programs require tablets). For anything you weren't able to check off with the first two, opt for a zero waste online store or general ethical online shop for eco-friendly back to school supplies as an opportunity to start teaching your kids about conscious consumerism.

For items you end up having to buy new, this list of the best eco-friendly school supplies (including sustainable stationery) will hopefully help. We've tried to find at least one solid sustainable item to fill each major common item. Most of us need office supplies at some point or another, and the two categories are really all the same. So no matter how old you are, there's just as much reason to make sure your office (and school) supplies are eco-friendly.

1. Plantable Pencils

Sustainable FSC certified wooden pencils are zero waste and have a non-toxic natural clay and graphite core. Then, instead of an eraser, they're capped with a biodegradable seed capsule. When your pencil gets too stubby to write with, just stick it in some soil as per their planting guide. Use the crafting potential of these to teach your child some apartment gardening basics! Choose from plain graphite, colored pencils, and inscribed sets. The Mindful Thoughts edition, which bears phrases like "All of us need to grow continuously in our lives", might even help get those creative juices flowing.

2. Recycled Newspaper Pencils

If you're worried about your kids losing their pencils, another eco-friendly pencil alternative is "tree-free" recycled newspaper pencils. These HB soft graphite pencils are comprised entirely of recycled newspapers and magazines. No wood at all, and they look pretty to boot. With an easy peel-to-sharpen design, these are great for kids. Just make sure to help them compost the peels to be truly zero waste. Sets of 5 or 10 come packaged in compostable, unbleached paperboard boxes.

3. Bamboo Pencil sharpener

Sharpen your eco-friendly pencils with this double hole eco-friendly pencil sharpener made of sustainably sourced bamboo and recycled stainless steel. At the end of its life, remove the blades for recycling and compost the bamboo body. It comes packaged with an unbleached cardboard backing printed with soy inks. The only slight downside is the recycled (and recyclable) plastic bit that holds the sharpener inside. This means it's not totally zero waste, but it was the closest we found on the market.

4. Natural Eraser

The dual-sided eraser made of all-natural rubber latex (to erase pencil) and natural silica sand (to erase ink and some markers). One zero waste eraser for all your mistakes. The individual erasers are packaged in a protective, 100% recycled pulp sleeve, which can be composted. Avoid the multi-packs because those are bound together in plastic packages.

5. Natural Grass Pen

Zero waste pens still leave us drawing a bit of a blank. While more sustainable pens exist now, they're still largely wasteful and greenwashing is still a concern, too. The truly best zero waste pens are refillable fountain pens. Great for the office, but perhaps less so for the classroom. That said, you can pick up a fountain pen for a reasonable price and while not totally zero waste either, we feel it's still one of the best alternatives. A good balance of very low waste while still being affordable. If you look after them they'll last a very long time. And you can sometimes find great secondhand fountain pens to reduce your waste further. Otherwise A Natural Grass Pen is a good option for kids. Just remember to gather up all those pens lying around the house and/or office, and use them first. When you're done so they can be recycled!

6. Organic Cotton Pencil Case

Now you need something to store all those eco-conscious goodies in, If you don't have a ditty bag lying around the house check out this line of eco-friendly school supplies from Canada. Their 100% organic cotton zippered pencil bags are available in tons of colors. Each bag is handwoven by underprivileged women in rural area in different countries, so as to provide jobs to these communities. The thick, ultra durable weave is designed to last, even when it gets buried under books in a backpack.

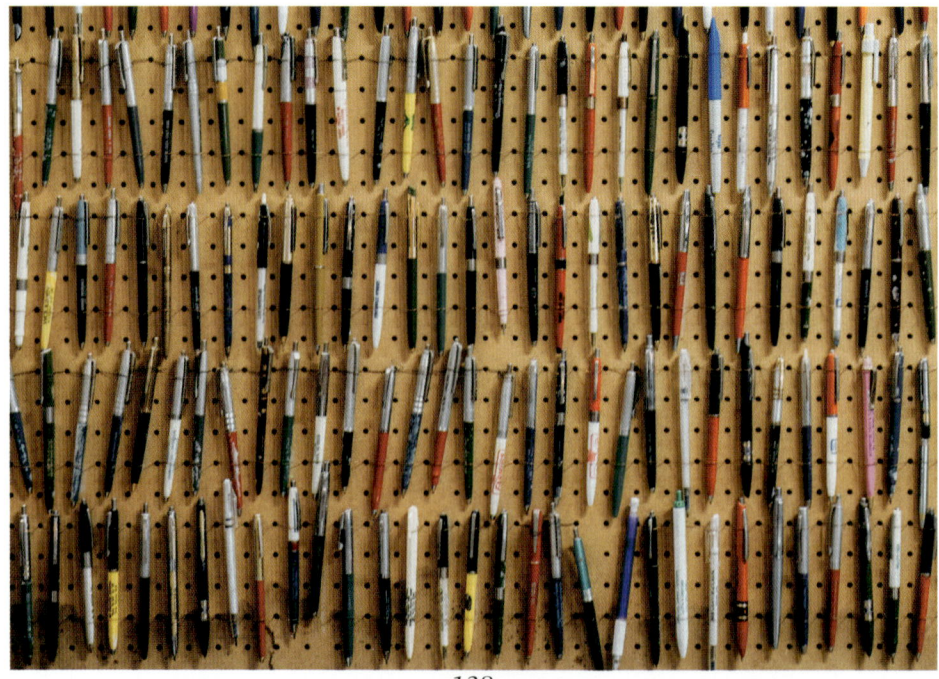

7. Decomposition Books

The best sustainable notebooks we found are different type of Decomposition books. These lovely college-ruled decomposition books are an excellent non-toxic and compostable note-taking solution. Each page is made of 100% post-consumer recycled paper. With so many nature-inspired cover designs (which are printed using soy inks), you can get something different for each subject or mood. Producer also makes spiral-bound notebooks, though we suggest avoiding if possible. Spiral bindings are inherently more wasteful.

8. Jotter Notebooks

Simple and elegant, these 30% recycled notebooks bound with wax thread are good for both school, work, sketching, journaling, or just making lists. Choose between blank, lined, or 3mm grid dot sheets (which would be perfect as graphic paper for math class). These zero waste notebooks measure 5" x 7" and come with a choice of either 30 or 60 sheets. You can also order them either individually or in packs of three or six. After you're done with the notebook, you can compost the pages and plant the seed paper cover to grow wildflowers! The embedded seeds will grow a blend of Snapdragon, Bird's Eye, Black-Eyed Susan, Clarkia, Sweet Alyssum, and Catchfly.

9. Recycled ReBinder

An undyed zero waste binder is never going to be as cool as the Trapper Keepers you coveted as a kid. But the planet will thank you! No more synthetic fabric-covered or plastic binders (one of many types of plastic, which are lucky to last even a semester). These recycled chipboard ReBinders are a far more durable choice, with a better end of life when they do wear out. Easily screw out the rings for recycling and throw the rest in the compost! And you could even draw on your own designs!

With 0.5" rings, you can fit about 100 sheets of recycled loose leaf paper. Fortunately, that's now pretty easy to find at just about any major office supply store.

10. 2-Pocket Folder

Unbleached folders are made using 22-point chipboard, which is 100% recycled FSC-certified post-consumer waste. This material is thick and durable, eliminating the need for any toxic coatings, including acids that can harm papers. Plus, then your kid (or you for that matter) can decorate it as they like. They're sold in sets of 25 to minimize shipping and packaging.

11. Recycled Tab Dividers

About Package Free Shop Recycled and Zero Waste Tab Dividers Your zero waste binder may help keep the planet clean, but how do you keep it clean and organized? With 3-ring compostable chipboard tab dividers, of course! These 5 or 8 tab sets are made out of 85% post-consumer and 15% post-industrial recycled content. With a 13-point thickness, they're durable and should last well beyond one school year. When they do wear out, you can home compost them.

12. Recycled Copy Paper

Not only is Printworks paper made of 100% recycled post-consumer waste (specifically from food and beverage containers), it's FSC-certified and chlorine-free. Unlike many recycled paper products that get shipped to Asia and back again, the waste is collected and remade entirely in the USA (only 300 miles from the mill, in fact). This reduces tons of carbon emissions from shipping. And since it comes in 20-pound boxes, you won't be needing to reorder anytime soon, either. It's just as affordable as traditional copy paper, too. We shouldn't have to choose between paying ourselves and making the planet pay.

13. Staple-Free Stapler

We don't have to prick our fingers pulling out staples before composting or recycling anymore? The PLUS Paper Clinch uses a unique folding technique to bind up to 10 sheets of paper just as staples would… but without all the waste and complication. It's also portable, kid-safe, and, according to the reviews, easy to use due to the ergonomic design and minimal force required. Unfortunately, the body is plastic,but if you take care to make it last, you're at least saving the staples and making whatever you're stapling easier to compost!

14. Paper Tape Cuts

Lasting Things paper tape cuts are made from upcycled 1970s vintage military surplus kraft paper combined with all-natural, compostable acacia gum. To activate the adhesive, just moisten a little bit and stick where desired. While this brown tape may not have all the advantages of clear tape, the biodegradable cello tapes can't be home composted. Besides, the kraft paper look gives it all sorts of crafty potential especially if used in combination with zero waste gift wrapping. An order includes 97 cuts of 9" x 2.5" sizes.

15. Eco-Friendly Backpacks

What good are eco friendly school supplies without a way to tote them? When it comes to something that takes quite a beating like a backpack, true quality is what you want, even more than sustainable materials. United By Blue offers tons of affordable eco friendly backpacks that are both. Their also are made of natural materials like organic waxed cotton canvas, or vegetable dyed repreve recycled polyester. With double stitching and DWR finishes, these are designed to last. They have many designs and sizes, many with internal laptop sleeves to function as two items in one.

16. Cork Laptop Sleeve

This minimalist, zero waste and vegan laptop sleeve is a great way to protect your hardware in your bag or on your commute. The cork provides some padding and water resistance while the cotton liner protects from scratches. But what is cork fabric? It happens to be one of the world's most sustainable materials because it's harvested by shaving the bark of cork trees, as opposed to cutting them down. This process can be repeated every nine years when the bark fully regrows, for up to 300 years. At its eventual end of life, just cut out the metal snaps and compost everything else. The many different sizes available ensure you can find something for every shape and size of laptop.

17. Newspaper Colored Pencils

School isn't all reading, writing, and arithmetic. Sometimes it's fun and we want sustainable art supplies to keep that fun, clean and healthy. Recycled newspaper makes for the best eco-friendly pencils (colored or not) which are great for two reasons: 1) they reduce landfill waste by putting newspaper to a second use; 2) they don't promote deforestation.

The tightly coiled newspaper held together by earth safe adhesive is sturdy to hold, while the non-toxic colored graphite goes on smooth and extra dark (which they claim provides double the life of traditional colored pencils).

18. Zero Waste Crayons

Conventional crayons are made of petroleum-based paraffin and are tinted with chemical dyes. To that, we say 'time to color outside the lines'. Earth Grown crayons are made of organic, vegan soy wax and mineral pigments that have been certified non-toxic by related third party organizations. All ingredients are sourced from farms and have been grown without pesticides or herbicides. They're also packaged in an uncoated cardboard box with shredded cardboard filling which is 100% compostable.

19. Art Drawing Pad

The Eco Art drawing pad is comprised of FSC and Rainforest Alliance-certified post-consumer agricultural waste. Specifically, it's made from pinzotes, the discarded stalks from banana trees. Each sheet is solid enough to be suitable for paints, markers, and crayons alike.

Environmental Shopping Guide for Paper Products

Sustainable purchasing is about including social, environmental, financial and performance factors in a systematic way. It involves thinking about the reasons for using the product (the service) and assessing how these services could be best met. If a product is needed, sustainable purchasing involves considering how products are made, what they are made of, where they come from and how they will be used and disposed.

Wherever possible RECOMENDED products that employ a combination of characteristics listed in the left hand column, and NOT-RECOMENDED products that demonstrate characteristic in the right-hand column.

RECOMENDED	NOT-RECOMENDED
• International Ecolabel certified • High post-consumer recycled fibre content • Non-wood • Chlorine free • Paper that is less bright	• Not Ecolabel certified • Long distance transport • Unsustainably harvested wood resources

A Simple Sustainable Plan

A	Reduce paper use
B	Use EcoLabel Certified Products
C	Choose High Post-Consumer Recycled Content
D	Choose Non-Wood (Tree-free) Fibres
E	Choose Less Transportation
F	Use Sustainably Harvested Wood Fibre
G	Choose Chlorine-Free Paper
H	Select Paper with Appropriate Brightness

A: Reduce paper use

The first step toward a more sustainable paper cycle is to reduce use. Using less paper saves money and contributes to sustainability by mitigating the environmental impact of production and use. The average North American now uses 227 Kilograms of paper per year, more than double the global average. Consuming less paper reduces the impacts of paper production and the associated energy use from operating printers and copiers. Both of these processes are known to have negative impacts for sustainability and should be minimized.

> **Strategies for reducing paper use include:**
> - Electronically archiving instead of printing non-critical documents
> - Sharing and reviewing document drafts electronically
> - Purchasing a duplex printer/photocopier and selecting double-sided printing as the default
> - Re-using paper that is already printed on one side for draft copies

Even if paper use declines in industrialized countries, developing nations will continue to increase their consumption as they gain access to more information and technology. Many countries currently have insufficient paper to fulfill basic education needs. It is therefore essential that paper consumption be closely monitored and restrained so that the resource, and the benefits it provides, can be more equally distributed to meet future needs.

B: Use EcoLabel Certified Products

Environmental Choice certified (eg. FSC certified International Environmental Labelling Series,Vol.6, Page: 29) paper products have met the required standards regarding noxious emissions to water, wastewater discharge levels, use of recycled content, solid waste volume, potential contribution to acid rain and climate change, energy use and forestry and habitat conservation. This widely respected sustainability rating provides an easy way of distinguishing genuinely green products from their competitors.

C: Choose High Post-Consumer Recycled Content

As opposed to virgin fibres, post-consumer recycled fibres have been recovered from paper products already "consumed" by an end user. The use of these recycled fibres directly reduces the use of forest resources, in turn mitigating the associated habitat destruction, loss of top soil and other forms of ecosystem damage.

> **Recycling one tonne of paper:**
>
> • Saves up to 31 trees, 4,000 kWh of energy, 1.7 barrels (270 litres) of oil, 10.2 million Btu's of energy, 26,000 litres of water and 3.5 cubic metres of landfill space
>
> • Burning that same tonne of paper would generate about 750 kilograms of carbon dioxide
>
> • Recycling paper saves 65% of the energy needed to make new paper and also reduces water pollution by 35% and air pollution by 74%

D: Choose Non-Wood (Tree-free) Fibres

Non-wood plant fibres do not need to be bleached with chlorine to be lightened, consume less energy when being processed, release fewer greenhouse gases in the production process and has less harmful water discharge. Crops grown specifically for the purpose of paper production can include Kanaf, jute, flax and hemp. Certain agricultural residues, such as wheat stalks and sugar cane bagasse, can also be processed into non-wood paper. When choosing non-wood fibres, preference should be given to those that are organically and sustainably grown. This eliminates the use of synthetic fertilizers, herbicides and pesticides, reducing the associated ecological and human health impacts.

E: Choose Less Transportation

The proximity of where the fibres are harvested, where the final products are produced and your own location has a significant impact on the sustainability of paper production. Growing, producing and buying locally will reduce emissions from fossil fuels. Transportation may have to be weighed against some of the other desired characteristics. More information regarding sustainable methods of transportation is available in the Transportation Guide.

F: Use Sustainably Harvested Wood Fibre

When using wood-based fibres, it is important to consider how the forest resources from which the paper was derived are managed and harvested. Preference should be given to companies that practice sustainable forestry techniques. Third party organizations such as the Forest Stewardship Council (FSC, Vol.6 Page 29) certify the harvesting and management of forestry resources to ensure long-term sustainability.

G: Choose Chlorine-Free Paper

To reduce the potential risks associated with chlorine compounds, a number of paper manufacturers are switching to chlorine-free compounds for whitening paper. Alternative bleaching agents include: oxygen, hydrogen peroxide or ozone treatments. Paper products often identify the bleaching method used for processing pulp. Paper products processed with derivatives of chlorine produce fewer dioxins than regular chlorine. This process is described as elemental chlorine free (ECF). Products bleached with no chlorine and no chlorine derivatives are sometimes referred to as totally chlorine free (TCF) or process chlorine free.

H: Select Paper with Appropriate Brightness

The brightness of paper is largely a function of the chemicals used in the pulp and/or the amount of recycled fibres used in the paper. Selecting less bright paper can reduce overall impacts. The function or use of the paper influences how bright the paper needs to be. For example, copy paper is generally not used for publicity or advertising. When the content, rather than the appearance of the paper matters, the whiteness is irrelevant as long as the text is legible. It is important to appropriately match the paper to its purpose.

How can we make stationery eco friendly?

Making your supplies & stationery stockpile more sustainable

- Make sure your paper has been sourced sustainably.
- Print double-sided and black & white.
- Cork noticeboards.
- Swap pens for pencils where possible.
- Reuse and recycle boxes and packaging.

Step Seven

Do It Yourself

STEP SEVEN

All about 'Eco-friendly' DIY Building, Modifying, and/or Reconstruction

DIY is short for do-it-yourself. It means carrying out home repairs, maintenance, and improvements yourself instead of hiring a professional. Interest in DIY took off after the Second World War. Changes such as growth in home ownership and the arrival of TV programs about home improvement helped to fuel the DIY movement.

After a while our homes need a change or some updating. They can seem tired and old and in need of a little refreshing. Ideally, perhaps, you might like to move to a larger home but the economy has you worried and you want to spend your money wisely. So instead, you've decided to renovate your home not only to perk it up but to better accommodate your current needs and lifestyle. If you haven't thought to do so already, you might want to think about some environmentally savvy ways to renovate your home.

Top Ten Eco-Friendly Ways to Renovate Your Home

If you are planning on renovating your home, you'll want to make sure you do it in style but without further causing harm to our environment. Contrary to what many people think, there are plenty of ways to make your home renovation an environment-friendly one. That way, you have a home that is not only aesthetically pleasing, but one that reduces environmental impact. But how exactly do you renovate a home to make it simultaneously eco-friendly and stylish? we show you Top Ten of eco-friendly ways to renovate your home:

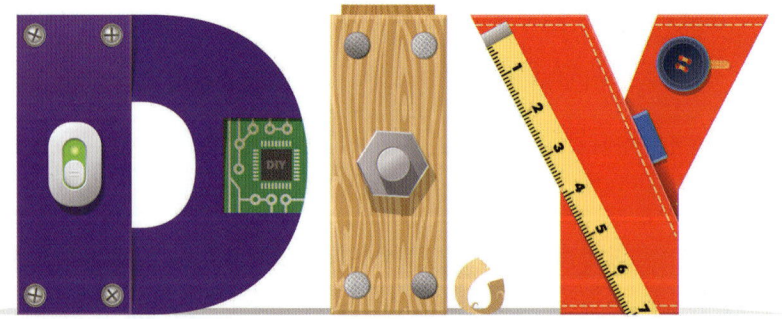

1. Use recycled glass

There are many home depots that now cell bio-glass that look like new windows but are 100% environment-friendly. They make a fantastic addition to the kitchen or your bedroom, as it brightens the whole space allowing natural lighting and morning sunshine – a great way to start your morning!

2. Use formaldehyde-free cabinets

Formaldehyde is commonly used in building materials and household products. But even if it's found in many household and beauty products, it's actually quite toxic! We recommend that you look for those free of this arterial to ensure safety for both your home and the environment's well-being. There are now many stores that offer VOC and formaldehyde-free furniture such as kitchen cabinets.

3. Paint with low-VOC or VOC-free paint

VOC is short for volatile organic compounds which produce harmful molecules. This ends up with you and the household incurring long-term health effects, and they aren't good ones either! Using VOC-free products like paint will help you breathe easier when at home.

4. Go solar

The sun is very powerful and it's a renewable source of energy – so use it to your advantage! Collecting the power of the sun via solar panels can give you electricity to last the whole night and you can use it to heart's content completely guilt-free! Not only will you help the environment through saving energy, but you will notice your electricity bills becoming less expensive as well. (For complete detail in this regard refer to IEL Vol.2 Energy)

5. Deconstruct your home – don't demolish it!

If you plan on tearing down walls or even knocking down entire rooms, walk around your home first to see what you can salvage and re-use beforehand. Not only is this eco-friendly, but it will save money in the end. If it ain't broke, don't fix it! Most likely there is a ton of material you can salvage and re-use. Consider everything from light fixtures, to flooring, tile, bricks, cabinets and molding. If you plan on replacing the chandelier in your dining room, instead of tossing it, think about using it in another room – maybe your kitchen, your daughter's bedroom, even a bathroom!

6. Choose bamboo flooring

What makes bamboo different from other types of wood materials? It's durable, moisture-resistant, grows back faster than wood and growing it uses less pesticides. We have an abundance of bamboo that can be harvested without destroying its roots, making it an environmentally friendly option. You get to save the lives of other trees and old growth forests, all while getting sleek chic flooring. Just make sure to get a good hammer to carry out this project; plus, it's energy-saving as well!

7. Using salvaged wood or discarded metal & Donate your unwanted items

Using recycled materials such as wood and metal will help to reduce waste and the need for fossil fuels as trucks and machinery aren't required to cut down existing trees. Plus, it gives your home a unique and contemporary look. Don't worry about rust or damage, as there are many types of salvaged materials that hold durability and beauty. So you really don't want that dining room chandelier in any other room. So you really don't want that dining room chandelier in any other room. Don't toss it, Perhaps you even have a crafty friend who might enjoy repainting and re-purposing it. With this in mind, not only are you being environmentally friendly, but you are truly giving back to the community.

8. Focus on energy-efficient appliances

If you're planning to replace refrigerators, air conditioners, or other types of major appliances, focus on those that help save both the environment and energy. Investing in energy efficient white goods will go a long way to saving you money in the longer term! (For complete detail in this regard refer to IEL Vol.2 Energy)

9. Consider buying pre-owned materials

Habitat for Humanity is one such retailer, but there are many across the country, some even specialize in high-end products. This can be a great and cost-effective way to redo your home. If, in the end, that SubZero fridge is an absolute must have, you could save thousands of dollars buying one that has been used for a couple of years. Cabinets may be the largest expense of a kitchen renovation, these salvage shops often have high quality cabinets in fabulous condition. It's an idea certainly worth investigating.

10. Re-face instead of replace

As I just mentioned, the greatest expense of any kitchen remodel may very well be your cabinets. Instead of replacing them altogether, consider repainting them or simply refacing them. Most likely your cabinets are in great conditions. New doors and drawers can give seemingly tired cabinets a whole new life!

Conclusion:

Looking for a way to renovate your home while saving the environment isn't all that difficult if you follow these eco-friendly renovation tips. And these helpful hints don't just apply to your home – they can apply to your office, work space or any other building project you're thinking of undertaking!

Do you have any other tips or suggestions on how to renovate your home in a more sustainable way? Please let us know for considering in the next edition of this Book.

Most Popular Eco-Friendly Flooring Solutions

We have provided a guide of the most popular eco-flooring solutions, some are new, some are old and a few will make you think:

1. Cork
Cork flooring is a product made from the bark of the cork oak tree, a material which is ground, processed into sheets and baked in a kiln to produce tiles that serve as flooring for offices, light commercial locations, and residences. Cork is harvested from the bark of the cork oak tree commonly found in the forests of the Mediterranean. The trees are not cut down to harvest the bark, which will grow back every three years, making it an ideal renewable source. It has anti-microbial properties that reduce allergens in the home, is fire retardant, easy to maintain and acts as a natural insect repellent too.

2- Bamboo

Bamboo flooring is another wood like option that is gaining in popularity. It is actually a grass that shares similar characteristics as hardwood. It is durable, easy to maintain and is easy to install. Bamboo is sustainable and made from natural vegetation that grows to maturity in three to five years, far less than the twenty years trees can take.

Bamboo, while usually very light, is available in many hues that will work in any setting or decor. Its varied grains and wide array of colors give it an edge over traditional flooring by allowing for customization not often found elsewhere.

FLOOR AND MATERIAL ICONS

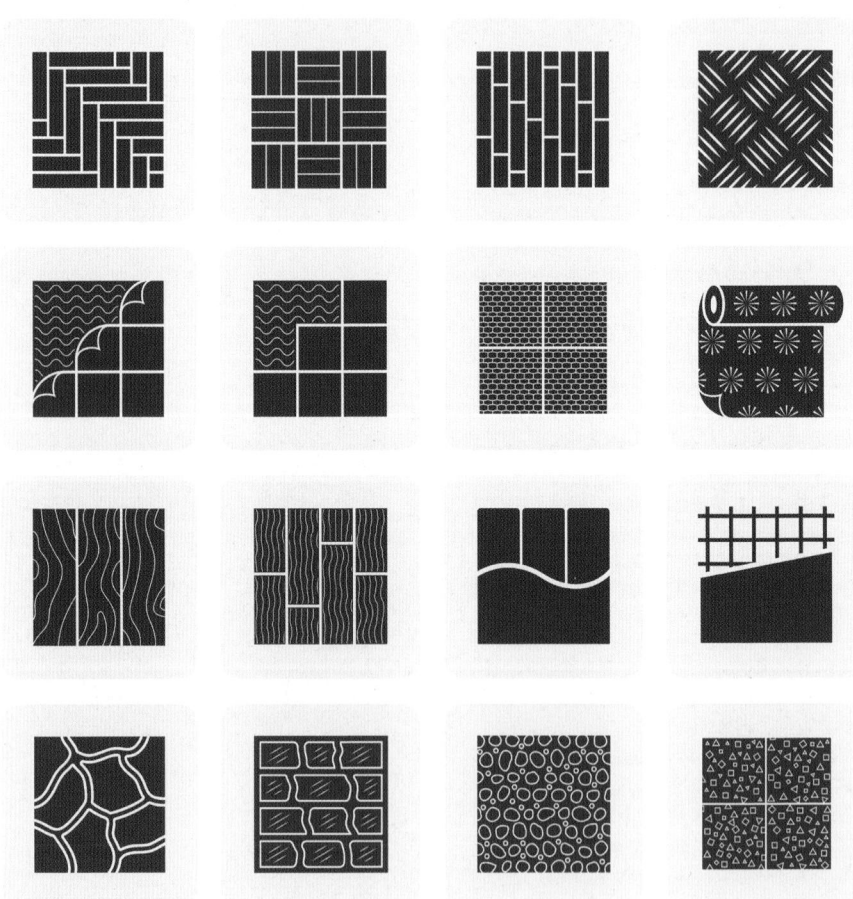

3. Linoleum

Linoleum is one of the most natural and sustainable flooring solutions on the market, appreciated for its natural beauty, comfort and durability for over 150 years. When one thinks of linoleum flooring, vinyl tends to come to mind and yet the two are nowhere close to each other. Vinyl is a synthetic made of chlorinated petrochemicals that are harmful. Linoleum is created from a concoction of linseed oil, cork dust, tree resins, wood flour, pigments and ground limestone.

Like cork, it is fire retardant and water resistant. Linoleum is not new to the market; it fell out of favor with the introduction of vinyl in the 1940's. As architects and designers began asking for it again, it reemerged with a vast array of bright vibrant colors and a new sealer to protect it from stains. It has a long shelf life and will hold up to a lot of wear and tear. In addition to being made with renewable materials, linoleum is biodegradable and won't take up space in landfills. Linoleum does not emit harmful VOCs (brand new linoleum does have a harmless odor from the linseed oil content that dissipates after a few weeks).

4. Glass tiles

Ever wonder what happens to the beverage bottles that are shipped to the recycler? They are converted into beautiful glass tiles. This renewable source is fast becoming a wonderful option for floors as well as bathroom and kitchen walls. Glass has similar benefits of other eco-friendly materials. It is non-absorptive and won't mildew or mold in damp environments. It is easy to maintain and won't stain.

Glass comes in a limitless array of colors, patterns and finishes suitable for most design schemes. Unlike ceramic tiles, glass will reflect light rather than absorb it, adding that additional layer of light some rooms need.

INTERNATIONAL MANUAL OF CLIMATE CHANGE CONTROL (IMCCC) • 167

5. Concrete

Polished concrete is an unlikely sustainable material that is gaining in popularity. Concrete is typically slab on grade and used as a sub flooring in some residential settings. If it is polished and tinted to the homeowners taste and style there is no need for traditional flooring to be put over it.

From creating a tiled effect with different colors to inlaying other materials such as glass the design possibilities are endless. Concrete is extremely durable, easy to clean and never needs to be replaced.

CONCRETE SLAB ICON SETS.

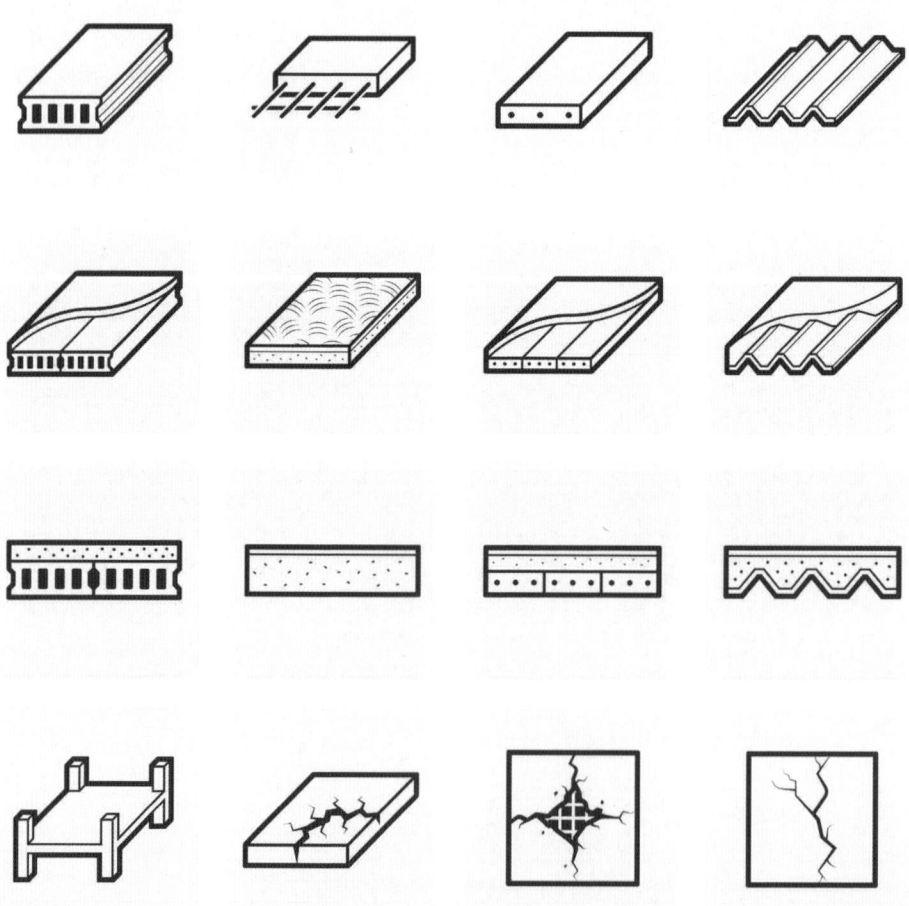

6. Organic wool carpet
Carpet has long been a favorite go-to material for most homes. It is soft to walk on, comfortable to sit on and comes in a range of colors and patterns. Unfortunately, carpet has typically been made using volatile chemical compounds or toxins that are harmful to the environment and to our health. There are eco-friendly options though.

Consider carpets made of organic wool. Organic wool is a natural resource spun into a thread that can be dyed any color imaginable, and then be woven to create a carpet. It is one of the first materials to be used as a floor covering, is very durable and can last centuries. In some families wool rugs have been passed down from generation to generation making them family heirlooms. At the end of its useful life, the pile from wool carpet can be returned to the ground, where the nutrients released as it decomposes promote further grass growth, and the natural production cycle starts all over again. Under the right conditions, wool is totally biodegradable.

7. Corn Carpet

One of the newest environmentally fibers, Carpet is made of corn sugar. It is a very high performing carpet fiber. It is one of the best in terms of durability and the best in terms of stain resistance (in our opinion). Since utilizing corn sugar may take away from the animal and human feed, this may make it not as environmentally friendly as the previous two options. However, you could argue that the durability of Sorona Smartstrand makes it more eco-friendly than PET polyester. It depends on how you look at it. This is a very exciting new carpet fiber.

Corn carpets are made of 35% renewable materials. What are "renewable materials"? It typically refers to a plant because they can be re-grown. In this case, corn sugar is the renewable component. These carpets are made from propanediol, or Bio-PDO, a corn-based polymer. It is made from corn sugar, a by-product of making ethanol, generated at a nearby plant.

An E. coli bacteria --genetically-modified by DuPont scientists--breaks down the corn sugar through a fermentation process that is much like making beer.

DIY
How to make homemade natural wall coating from recycled clothes and fabrics

There are many reasons to use natural wall coating on your walls; like it being odor-free, sound insulation, moisture insulation, heat and cold insulation, crack resistant and quick and easy applicability. This natural coating can cover glass, ceramic, cement, brick, wood. It also has color diversity, acoustic insulation, high resistance against humidity, applicable on any surface, odore free, repairable, 100 % natural, clean environment, thermal Insulation, fire resistance, no more insect, light and many other benefits.

Recipes for homemade natural wall coating:

If you're wanting to ditch those toxic, commercial chemical paints for coating walls and switch to a more natural, homemade wall coating: this simple recipe will have you coating green in no time. However, before we get to the wall coating, let's check out some of the most common (and most useful) non-toxic products:

Small Shredder

Firstly, you need to purchase a $50 DIY shredder to crush recycling (if your paper shredder is strong enough for fabrics and used clothes, you can use it). It will be around $50 only if you have access to a waterjet cutter, but you most likely don't and the minimum will round to about $150. You can have your home shredder for textile, plastic and metal.

Necessary items for start wall coating
- Recycled Fabrics and clothes
- Small home Yelca Shredder
- Added to Yelca powder and required water
- A plastic pan or basin in which shredded clothes, Yelca powder and water can be mixed.
- A medium or large size Yelca trowel or plastic trowel for rubbing and smoothing Yelca paste on the desired surface.

How to prepare Yelca dough:
Do the following steps in order and carefully:
Step 1: Collect used fabrics and clothes for as much that is needed
Step 2: Give it to the small home Yelca Shredder and start to shredding above mentioned fabrics and clothes
Step 3: Add required amount of Yelca powder
Step 4: Make and mix right amount of water with items mentioned in Step 1 and 2 in the existing plastic pan.
Step 5: After 15-30 minutes the dough is ready to be installed on the desired surface.
Step 6: The correct way to rub Yelca on the desired surface is to: first, thoroughly clean the surface of the wall to remove soot, dust or any obstacles to the adhesion of Yelca paste. Then, with a special Yelca trowel, spread the Yelca dough on the surface until it reaches a thickness of about one millimeter.
Step 7: Be careful in doing this and do not give up quickly because you may not be able to do it well the first time, however, with practice and repetition you can achieve a skill that covers both the entire surface with the dough and the thickness of the dough is minimal (not more than 0.5 mm to 1.5 mm).

> **Note:** We've done our absolute best to provide the best information possible, but since we haven't tried every single one of these DIY items in every possible situation, we can't vouch for them 100 percent. So please be cautious and take many safety pre-cautions.

For all People who wish to take care of Climate Change

For all Schools, Libraries, Homes and/or Offices

Available in more than 39,000 booksellers worldwide, including Amazon, Barnes & Noble, Google Play Books, Walmart, …. and many more.

The greatest expense of any kitchen remodel may very well be your cabinets. Instead of replacing them altogether, consider repainting them or simply refacing them. Most likely your cabinets are in great conditions. New doors and drawers can give seemingly tired cabinets a whole new life!

Step Eight

Agriculture & Gardening

STEP EIGHT

All about 'Eco-friendly' garden

The environment is – rightly so – on everyone's minds, and if you have a garden, you have an opportunity to contribute to protecting the natural world.

But what does a eco-friendly garden actually look like, and what features does it have? You might imagine an untamed, overgrown jungle teeming with wildflowers and insects.

Nowadays, however, even the slickest contemporary garden designs can be environmentally friendly, thanks to a combination of ethically-sourced materials and innovative technology.

Plan your eco-friendly garden today:

Seven ways to create an eco-friendly garden:

1. Recycle and Reuse Materials

Using recycled materials instead is a great way to go green. The main concern is the origin, extraction, manufacture and installation of materials in structures, paths, walls and patios.

As reclamation yards, especially those in cities, can be expensive, trawl through out-of-town yards and junk shops for materials. New' materials such as paving made from recycled concrete aggregate are now widely available and Many companies now sell pots, fencing and furniture made from recycled wood and plastic.

2. Looking for Ecolabelled Products

Green materials sourced and made by the local community feature strongly in sustainable gardens. Choosing them helps to reduce your carbon footprint as they have few air miles attached, plus most of them use little or no cement, the production of which accounts for more than five per cent of the world's carbon emissions. (Page 36 for more detail). They also give gardens a 'sense of place' by linking them to the local surroundings, which is especially important in rural settings. Materials such as cob (clay and straw), oak, rammed earth, log walls, woven willow, chestnut paling timber and even straw bales are full of character. You will need to consider cost versus durability more keenly than usual, but the suppliers and craftsmen will be able to advise you.

3. Use Local Materials
Choose materials and features, such as paving and pergolas, that have been sourced or built locally, as this will help reduce a garden's carbon footprint and support nearby businesses. For timber products and decking, look for a Forest Stewardship Council (FSC) logo, (Page 32 for more detail) for wood originating from certified plantations.

4. Reduce Water Consumption

Water conservation is essential, so install a butt on every downpipe – you can choose weathered oak barrels or the ubiquitous green plastic tubs. If you have the space, consider an underground rain tank. Larger ones can easily collect enough water for the average garden, plus you can set them up to flush your WC.

Clever cost-cutting irrigation helps too. Don't use a sprinkler on the garden – water the roots of plants without wasting it on the leaves (automated watering systems are useful here); repair leaky pond liners; buy large pots for plants as they don't dry out as quickly; and don't mow the lawn too low in hot weather.

5. Use Premeable Paving

Water run-off from concrete-covered gardens in towns and cities causes localised flooding and affects wildlife significantly. To tackle this problem, legislation has been introduced to regulate the use of solid surfaces in front gardens. You must now use permeable surfacing materials.

Crunchy gravel and slate chippings are the obvious permeable alternative to solid paving, but there are lots of other materials available – from porous asphalt and block paving to grass reinforced with recycled plastic grids.

6. Using Eco Roof

Green roofs are becoming increasingly popular as they help increase biodiversity, provide good insulation, improve air quality and control water run-off – they're also very attractive.

There are plenty of products available using different construction techniques – you can even retrofit an existing shed and garage if they are able to take the weight.

7. Choose Eco-friendly Plants

Choosing the best plants is an important design tool, especially if you're looking to create an organic garden.

In an Eco-friendly garden, the best plants will provide food and shelter, creating perfect habitats for beneficial wildlife. Choose lots of local berry-producing plants and trees, such as hawthorn, which might be growing nearby – birds and insects will already be used to them, so they'll visit your garden more frequently if you grow them.

5 supplementary ways to create an Eco-friendly garden:

1. Increase wildlife and biodiversity

Encouraging wildlife makes your garden far more entertaining, as well as helping with pest control – slug-eating hedgehogs and slow-worms love piles of leaves and logs. To attract birds that help with caterpillar control, erect nest boxes and put out a variety of food. Song thrushes love dried fruit, blackbirds adore rotten apples, and sunflower seeds will attract chaffinches and blue tits.

Entice bees by choosing plants with 'open faces'; bright, showy blooms that flower throughout the year. Spring flowers are important for waking bees but are often overlooked; wallflowers, aubrietia and rosemary are all good pollen and nectar sources. For summer, catmint, thyme and lavender are particularly good. Ivy in flower is great in autumn. To help solitary bees, a special bee hotel provides the perfect nesting site.

The key principle to planting is always putting them where they're happiest. Contented plants take care of themselves, but stressed ones need constant feeding and watering, so make sure you don't plant your sun-lovers in the shade, for example, or vice versa. Matching the right plant to the right place will also help keep garden maintenance time to a minimum.

2. Change quality of Soil

Lots of compost and/or well-rotted manure will keep your soil in what gardeners call 'good heart'. This creates a healthy soil teeming with essential micro-organisms, which in turn gives you healthy plants that don't succumb to pests and diseases. Compost soaks up water like a sponge, too – useful in free-draining sandy soils.

Dig in a large bucketful every few feet when planting, or spread liberally around plants as a mulch each spring; this also helps stop light soils from being washed away in heavy rain.

3. Prepare homemade compost

Recycling green waste is important, too – homemade compost costs nothing to make, and will save money on bagged compost and soil conditioner from garden centres.

What to add to your compost:
- veg peelingswood
- hedge trimmings
- grass clippings
- tea bags
- egg boxes
- egg shells
- leaves
- shredded paper
- vacuum cleaner contents

Avoid cooked food, meat, pet faeces and glossy magazines

4. Limit peat-based composts

Peat is dug from peat bogs, causing irreparable damage to precious natural habitats, and as it forms very slowly, simply isn't sustainable either. The problem is that it's great for growing plants as it's sterile, easy to handle, and holds onto nutrients and water like a sponge.

To cut down your use, stop using peat as a mulch or soil conditioner, and instead use **homemade compost**, rotted farmyard manure and leaf mould, which perform better and are full of nutrients. Buy peat-free or reduced-peat compost for potting – it'll say so on the bag – and you'll now find that modern peat-based composts available to gardeners include some of the alternatives, such as bark, wood fibre, coir (coconut husk) or specially formulated green waste.

5. Create a natural garden

Plant according to the garden, not the gardener,' is the ethos at the heart of the natural garden, and looking to nature will provide inspiration and a template to follow. Work with the characteristics of your garden, not against them.

For example, in damp shade, embrace woodland plants and those that grow on woodland margins. For sunny slopes, consider Mediterranean plants like rosemary, juniper, bay and sage – plants with silvery or blue-grey leaves that have naturally adapted to such conditions. Waterlogged soil? Choose wetland plants – not only will your planting visually sit more comfortably, but also promote happy, healthy plants and suffer fewer pests and diseases.

More popular now than ever, native plants are tough, easy to grow and provide food and valuable habitats for wildlife. Ideal for a more relaxed design, they will also help to preserve our threatened plant heritage. Favourites include spiky teasels, gunmetal-coloured cotton thistle, and hardy cranesbills. Quick to colonise poor soils and sunny walls, the red, pink or white flowers of valerian, in particular, are stunning and last for ages. Leave some parts of the garden untidy; nature likes it messy, so gather piles of leaves in undisturbed corners and collect logs and branches, rather than burn them (if they're not diseased). You'll encourage thousands of insects and foraging birds. Hedgehogs also find such spots irresistible to hibernate in.

All about 'Eco-friendly' Livestock Ranching

Ranching is the act of running a ranch, which is essentially an extensive farm for the sole purpose of raising livestock and crops. ... Therefore, we can bring the two definitions together to define livestock ranching as the breeding of animals, for the purpose of food or clothing production. livestock farming, raising of animals for use or for pleasure. ... Ruminant (cud-chewing) animals such as cattle, sheep, and goats convert large quantities of pasture forage, harvested roughage, or by-product feeds, as well as nonprotein nitrogen such as urea, into meat, milk, and wool.

Livestock farmers are facing a number of challenges today. Demands for a lower impact on the environment, especially reducing greenhouse gas emissions, for more animal welfare and for less intensive production need to be balanced with a stable production and a good income. While there are challenges for livestock farmers, there are also many opportunities to increase the resilience and profitability of their farms.

Sustainable pasture management can offer ways to provide good feed to dairy cattle, help reduce livestock emissions, build up and store carbon in the soil, help mitigate the effects of climate change, and much more.

Digitisation and decision support tools can support farmers in better managing their livestock, for instance for more resource- and cost efficiency.

Grazing for carbon

Grasslands have enormous potential for storing carbon (C) in the soil. Carbon sequestration improves soil health, makes soils more resilient to extreme weather events, contributes to climate change mitigation and can benefit pasture quality. In sustainable livestock grazing systems, the key challenge is to find the best type of management to combine animal production with soil ecosystem services such as carbon storage, nutrient cycling and biodiversity.

Marketing pasture while reducing emissions

Livestock production significantly contributes to ammonia and greenhouse gas emissions, specifically methane. Adopting grazing methods that let cows graze optimally can contribute directly to lowering livestock emissions.
In addition, there is a growing interest in precision livestock farming, measuring methods, and digital tools that can support farmers in lowering, monitoring or managing farm emissions.

The 'Grazing cow monitor' project has developed a collar that uses GPS tracking to monitor the location of individual cows. The tool clearly indicates how much time a cow spends indoors or outdoors. "The monitor gives farmers digital proof that their cows have spent a sufficient amount of time grazing outdoors", It can therefore help make their administrative tasks easier, but it also allows farmers to label and market their milk as pasture milk." Giving cows access to pastures provides them with fresh grass to eat and can help maintain a healthy soil ecosystem. Outdoor grazing can also help to reduce ammonia emissions from livestock.

When the cows spend less time in the stables, this lowers the chance of faeces reacting with urine and producing ammonia. Keeping track of their cows' pasture time is one of the options for farmers in Flanders and the Netherlands to prove that they are taking measures to lower ammonia emissions.

Farmers can see the results on a dashboard, which they can easily access on their computer or smartphone.

Strategies for managing permanent pasture
Permanent grasslands offer many benefits for biodiversity, ecosystem services such as carbon sequestration, and animal health. Sustainable management strategies can help to maximise these benefits, for instance by matching grassland production with livestock needs.
Digital measuring and decision support tools can help increase resource efficiency and optimise grass production. Differentiating grass-based products such as meat, milk and cheese, can help create higher market value for farmers. Exchanging knowledge is key for increasing profitability, productivity and sustainability for permanent pastures all over the globe..

Robust and resilient dairy farming

Dairy farms are currently faced with economic and environmental challenges, such as volatile prices, extreme weather events, and market demands for more animal-friendly production systems. Improving grazing management can lead to happier cows that produce quality milk with a better price for the farmer. Exchanging experiences can help farmers make their own farms more robust and resilient.

How to create eco friendly Butterfly Garden

Creating a hospitable environment will entice butterflies to stay around long enough to lay eggs for a new generation. By providing the basics of shelter, water, and food—including butterfly-friendly plants—butterflies have a greater chance of thriving and reproducing. Kids love butterflies! Encourage your child's sense of connection to the natural world and invite butterflies into your landscape by planting a butterfly garden. A butterfly garden provides a colorful array of nectar-producing plants that not only attract butterflies (and often hummingbirds as well), but offers plants to feed the caterpillar stage of their life cycle. With the appropriate plantings, a butterfly garden provides opportunities to educate your children about the life cycle of a butterfly, allowing them to view each stage of growth and explore the intricate relationships of plants and animals.

INTERNATIONAL MANUAL OF CLIMATE CHANGE CONTROL (IMCCC)

Materials:
- in-ground garden space, raised bed or container garden
- trowel or shovel
- flowering plants for adult butterfly
- host plans for caterpillars

Approximate Time to Complete: 3 to 4 hours to plan, gather plants and install; multiple weeks to grow and attract butterflies

Location: Outdoor
Ages: fun for all ages
Season: spring through fall

Start with a blank slate by pulling out dead bush, removing the plants that were overgrown, and just getting down to the bare earth.

Select plants that grow well in your area. You need to include flowering plants that attract butterflies (many butterflies have favorite plants to sip from) and also leafy "host plants" that attract egg-laying butterflies and provide food for the caterpillars (also known as the larvae). It is always best to select native plants that will attract butterflies native to your area.

Preferred Plants For Caterpillars
Note: Planting trees and taller bushes will help give your garden the sustainability it needs to feed your caterpillars and preserve the flowers for your butterflies.

Trees
Serviceberry, River Birch, Hackberry, Kousa Dogwood, Apple, Cherry, Plum, Willow

Here are a few examples of common butterflies
and their preferred food sources:

Butterfly	Host plant(s) for caterpillars	Nectar plants for adult butterflies
Monarch	Milkweed	Milkweed, asters, red clover, zinnia, cosmos, lantana, pentas, daisy
Eastern Black Swallowtail	Carrots, celery, dill, parsley, Queen Anne's lace, rue, Texas turpentine broom	Milkweed, phlox
Giant Swallowtail	Citrus, hop tree, prickly ash, rue	Lantana, orange tree
American Painted Lady	Daisies, everlastings, and other composites	Burdock, daisy, everlastings, mallow, yarrow, zinnia, heliotrope
Orange Sulphur	White clover, alfalfa, vetch, lupine	Clovers, dandelion, parsley, zinnia, other meadow flowers, members of the composite family
Silver-Spotted Skipper	Beans, beggar's tick, licorice, locusts, wisteria	Many garden and meadow flower
Variegated Fritillary	Violets, pansies, stonecrops, passionflowers	Meadow flowers, hibiscus, composite family

Plant your garden. Add one or two large, flat rocks in the sun so butterflies a place to bask when mornings are cool. Since butterflies cannot drink from open water, provide them with a "puddle" by filling a container, such as an old birdbath, with wet sand where they can perch and drink safely.

Once the garden is planted, stand back and wait for the butterflies to stop by. With a successful butterfly garden, your kids will be able to observe the developmental process of a butterfly.

The eggs soon hatch, and the larvae appear and eat the leafy growth of the host plant, eventually developing into full-grown caterpillars. Remember, you will need to tolerate some leaf damage from your very hungry garden guests.

Later, these caterpillars affix themselves to a twig or branch and form a chrysalis, entering the pupa stage. Within about two weeks, they metamorphose into butterflies and re-emerge. Avoid all pesticides. Butterflies are insects, so it makes sense that insecticides — even those labeled organic" — can harm them.

Step Nine

Professional

STEP NINE

How can Professionals be 'Eco-friendly'

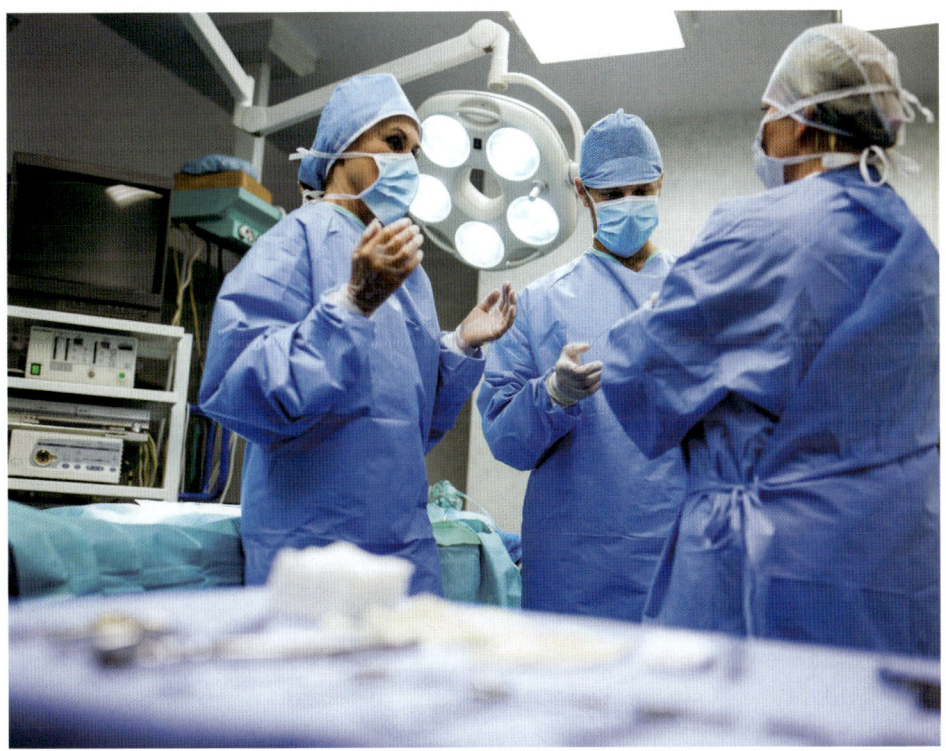

What is Being Environmentally Friendly? ... A good way would be to start with conserving water, driving less and walking more, consuming less energy, buying Ecolabelled and recycled products, eating locally grown vegetables, joining environmental groups to combat air pollution, creating less waste, planting more trees, and many more.

TEACHERS

How can teachers be eco friendly?

As a teacher, there are so many things you can do to make your classroom and school more green. From recycling, to planting gardens, to powering your school with solar panels to getting a green seal, the ideas are endless.

TTAIN has generated a list of 7 ways to go green in the classroom. This list covers everything from ideas for classroom décor to ways to make your classroom more energy efficient. Plus teaching students about green practices now creates a lifelong interest in saving the planet. These ideas help lay the groundwork for a green school and classroom. What will you start today?

1. Lead a green club

Find students who are interested in making their school more eco-friendly. Help them set a small goal to get started and then encourage them to think of more big picture ideas.

2. Prepare appropriate Text books

Textbooks are especially helpful for beginning Eco-friendly teachers. The material to be covered and the design of each lesson are carefully spelled out in detail. Eco-friendly Textbooks (such as **International Environmental Labelling Textbooks (Vol.1-11)** accompanied with a **knowledge test**, provide organized eco-friendly units of work. This series of textbooks gives you all the plans and lessons you need to cover to create a more green school environment. Order your package right now via (www.toptenaward.net).

3. Create a Recycling Center
A classroom recycling center is a great way to get kids excited about recycling. You can reduce the burden of sorting materials by designating a student "recycle ranger" to make sure everything is in its place.

4. Apply for grants
Find and apply for grants that offer financial support for green-school initiatives. There are lots of opportunities.

5. Green Mantras
Unify your classroom theme by adopting a green mantra. Display it prominently in your classroom. Ideas include: My choices make a difference. We have the power to make a difference in the world.

6. Eco-Friendly School Supplies
Promote using Eco friendly school supplies that are better for the environment.

7. Energy Efficiency in the Classroom
Turn off heating or cooling units and open the windows when the weather is nice. Enjoy the fresh air!

PILOTS

How can pilots be eco friendly?

Focusing on pilots' pre-flight, in-flight and post-flight behaviours and throwing in a few small incentives, the study led to some huge fuel savings and an unexpected jump in work satisfaction for many of the pilots. While commercial aviation accounts for 2.5 percent of global carbon emissions, the industry is taking strides to reduce its carbon footprint.

Eco friendly Pilots usually:

- Using less power for takeoff
- Taxiing with just one engine
- Maximizing cruising altitudes and winds
- More efficient circling
- Using less fuel to descend
- **Recommendation to all friends and family to have a complete set box of (International Environmental Labelling Vol.1-11) books in their home and/or offices)**
- Look at Environmental Labels

LAWYERS

How can lawyers be eco friendly?

While the world is still not far from the brink of irreversible damage from climate change, more and more companies–both big and small– are taking the necessary steps to become more sustainable. The same is true for law firms. As sustainability movements encourage people and businesses to go green, law firms are also stepping up to the plate. Here are some of the best ways law firms can achieve sustainability in their

offices:

- Reduce unnecessary paper use
- Choose sustainable office supplies (Refer to International Environmental Labelling Vol.6 wood and stationery)
- Minimize single-use cups and bottles (Refer to International Environmental Labelling Vol.1 Chapter 7)
- Minimize travel
- Invest in energy efficiency (Refer to International Environmental Labelling Vol.2 Energy)
- **Recommendation to all friends and family to have a complete set box of (International Environmental Labelling Vol.1-11) books in their home and/or offices)**
- Look at Environmental Labels

Advertising Professionals

How can advertising professionals be eco friendly?

Eco-friendly advertising or marketing refers to selling products or services based on their environmental benefits. These offerings may be environmentally friendly in themselves, or their production process is somehow ecologically responsible. Eco-friendly advertising campaigns highlight these benefits and share them with your consumers. Here are a few tips that will help you create a successful, eco-friendly advertising and marketing strategy:

- Focus on the Benefits
- **Recommendation to all friends and family to have a complete set box of (International Environmental Labelling Vol.1-11) books in their home and/or offices)**
- Think Locally
- Support Environmental Initiatives
- Always Be Transparent
- Look at Environmental Labels

ARCHITECTS

How can architects be eco friendly?

By using trees, plants, and grasses that are native to the area, architects can greatly reduce irrigation needs. Landscaping can also be used as part of a passive energy strategy. By planting trees that shade the roof and windows during the hottest time of the day, solar heat gain inside the building can be reduced.

Here are some of the best ways architects can achieve sustainability in their projects:
- Design in airtightness - Eco frienedly Design of buildings
- Use enough insulation - most buildings are built with too little
- Use the buildings thermal mass to best effect
- Choose the building materials with Ecolabels
- Make the best use of natural light
- Deploy renewable technologies only after your shell design is complete
- **Recommendation to all friends and family to have a complete set box of (International Environmental Labelling Vol.1-11) books in their home and/or offices)**

ACCOUNTANTS

How can accountants be eco friendly?

When it comes to the sustainable development of your accounting and bookkeeping business, one of the biggest things you should start with is reducing the amount of paper you use. Consider scanning your important documents and converting them into digital documents instead.

Green Accounting and Bookkeeping: 7 Tips for Your Business:

- Go Paperless
- Opt to Hold Virtual Meetings More Often
- Watch Your Electricity, gas, water Consumption for a planning reducing
- Use Recycled or Recyclable Office Supplies
- **Recommendation to all friends and family to have a complete set box of (International Environmental Labelling Vol.1-11) books in their home and/or offices)**
- Be Conscious of Your Transportation Choices
- Look at Environmental Labels

ENGINEERS

How can engineers be eco friendly?

Green engineering is the design, commercialization, and use of processes and products in a way that reduces pollution, promotes sustainability, and minimizes risk to human health and the environment without sacrificing economic viability and efficiency.

There's often an emphasis on businesses being environmentally responsible in their operations, but everyday professionals have a role to play too.

Here are the top ways that engineering professionals can be environmentally friendly:
- Avoid Paper Waste
- Get Educated Online
- Suggest Solutions at Work
- Use The Right Tech
- Choose the engineering materials with Ecolabels
- **Recommendation to all friends and family to have a complete set box of (International Environmental Labelling Vol.1-11) books in their home and/or offices)**
- Conserve Energy

CONSULTANTS

How can consultants be eco friendly?

An environmental consultant's goal is to help others make informed decisions about policies or projects that will impact the environment.

In short, they gather information, analyze it, and provide their recommendations.

These consultants might provide plans for reducing waste or conserving energy. An eco consultant might even produce a viable roadmap towards switching over to renewable energy. An eco consultant might look for more ethical sources of materials and offer guidance on sustainable purchasing practices.

Eco consultants may provide a wide range of services to help companies become more sustainable. According to Eco-officiency, these services mostly involve environmental assessments and plans for corrective measures.

An eco consultant might, for example, assess how a company utilizes natural resources like energy, water, and carbon, while also investigating how they dispose of waste products or hazardous materials.

Human Resources Specialist

How can HR specialist be eco friendly?

Environmental sustainability will only be achieved if we all do our part. And that includes those of us who work in the HR industry. Here are the top ways that HR professionals can be environmentally friendly:

- Cut out paper waste
- Invest in reusables to eliminate single-use plastics
- Initiate employee volunteer opportunities to create a positive difference
- Get creative with end of week waste
- Try to recommend people who think eco friendly for hiring
- **Recomendation to all friends and family to have a complete set box of (International Environmental Labelling Vol.1-11) books in their home and/or offices)**
- Conserve Energy
- Be Conscious of employee Transportation Choices
- Look at Environmental Labels

RESEARCH & DEVELOPMENT

How can R & D specialist be eco friendly?

R&D has an important role in improving the environmental performance of industry – an important element in sustainable development. International Energy Agency figures indicate that technologies and best practices could save between 17 and 27% of current primary energy use in global industry. Putting International R&D at the service of sustainable development is essential to our future.

Innovative production has a vital role in the quest for sustainable development. The ambition of eco-efficiency is to close the loops in the life cycle, so business models must adapt. And we must teach people how to work with people in other fields. Sustainable development is both a challenge and an opportunity for the process industries in the world.

Key aspects include:

- Implementing research for sustainable development
- Designing research policy for sustainable development
- Measuring the contribution of research to sustainable development

PSYCHOLOGIST

How can psychologist be eco friendly?

Conducting research on messages that motivate people to change their behavior. Spreading the word about environmental solutions. Uncovering why people may not adopt positive behaviors. Encouraging people to rethink their positions in the natural world.

Social psychology's contribution to a sustainable, flourishing future will come partly through its consciousness-transforming insights into adaptation and comparison. Conservation psychology is not only concerned with the ways psychology can contribute to protecting the natural environment, but also with how attention to the natural environment can contribute to psychology. ... It is well known, for example, that environmental toxins can have direct impacts on human health.

Here are the top ways that psychologists can act environmentally friendly:

- Effects on human behavior
- Infloencing on the public opinion about climate change, and
- Ways to modify the human sources of climate change

PHARMACIST

How can pharmacist be eco friendly?

Pharmacists can help to ensure that unused medications are returned to the pharmacy and disposed of appropriately, through hazardous waste companies. 3,4 By educating patients on proper disposal, pharmacists contribute significantly to preventing medications from entering the water supply.

Both recycling and waste reduction are important to making pharmacy practice more sustainable. Though patient education, recycling, and paperless communication methods are feasible short-term options, there are still a few barriers towards implementing these sustainable practices.

Here are the top ways that pharmacist can be environmentally friendly:
- Avoid Paper Waste
- Conserve energy (Refer to International Environmental Labelling Book series Vol.2 Energy)
- Encourage green logistics
- Sell eco-friendly products (Refer to International Environmental Labelling Book series Vol.4 and Vol.5)
- Order larger bottles
- Reuse pill containers
- Recycle (Refer to International Environmental Labelling Book series Vol.1 Chapter 7)
- **Recommendation to all friends and family to have a complete set box of (International Environmental Labelling Vol.1-11) books in their home and/or offices**

COMMERCIAL BANKERS

How can commercial banker be eco friendly?
By allowing card holders to use digital technologies to manage their finances, banks and other payment providers are providing a sustainable alternative to paper statements and physical bank branches. ... with innovative, more seamless experience, but it also allows banks to reduce their carbon footprint.

Here are the top ways that commercial bankers can act environmentally friendly:

- Moving away from paper
- Using sustainable materials and partnering with green suppliers
- Providing customer insight about their carbon footprint
- **Recommendation to all friends and family to have a complete set box of (International Environmental Labelling Vol.1-11) books in their home and/or offices)**
- Conserve Energy
- Encourage green logistics
- Choosing eco-friendly products

DOCTORS & NURSES

How can doctors and nurses be eco friendly?

Over the past seven years, hospitals are becoming more eco-friendly, seeking to lighten their environmental footprints. The benefits are tremendous! Among them are safer patients, less wastefulness, and lower facility operating costs.

Here are the top ways that doctors & nurses can be environmentally friendly:

- Avoid Paper Waste
- Conserve energy (Refer to International Environmental Labelling Book series Vol.2 Energy)
- Encourage green logistics
- Choosing eco-friendly products (Refer to International Environmental Labelling Book series Vol.4 and Vol.5)
- Order larger bottles
- Reuse pill containers
- Recycle (Refer to International Environmental Labelling Book series Vol.1 Chapter 7)
- **Recommendation to all friends and family to have a complete set box of (International Environmental Labelling Vol.1-11) books in their home and/or offices)**
- Reducing, treating, and safely disposing of waste

Here are the top ways that engineering professionals can be environmentally friendly:

Avoid Paper Waste
Get Educated Online
Suggest Solutions at Work
Use The Right Tech
Choose the engineering materials with Ecolabels
Conserve Energy

Step Ten

Financial

STEP TEN

All about 'Eco-friendly' Financial Products and/or Services

Green loans, energy efficiency mortgages, alternative energy venture capital, eco-savings deposits, and "green" credit cards; these items represent merely a handful of innovative, "green" financial products that are currently offered around the globe1 . In an age where environmental risks and opportunities abound, so too have the options for reconciling environmental matters with lending and financing arrangements.

The purpose of this Volume is to examine the currently available Eco friendly financial products and services, with a focus on lesson learning opportunities, the nature and transferability of best practices, and how key designs can potentially increase market share and generate profits, while improving brand recognition and enhancing reputation.

In this volume we have divides the financial services sector into the following categories:

1) Green Retail Banking;
2) Green Corporate and Investment Banking;
3) Insurance
4) Asset Management;

1. Green Retail Banking

Retail banking covers personal and business banking products and services designed for individuals, households and SMEs, rather than large corporate or institutional clients. Products and services in the eco friendly retail space include Green loans and mortgages, debit and credit card services, travelers' cheques, money orders, overdraft protection, cash management services and insurance, among others.

2. Green Corporate and Investment Banking

Corporate and investment banking, or "wholesale banking", sees banks provide banking solutions to large corporations, institutions, governments and other public entities with complex financial needs, typically international in scope. Financial institutions offering corporate and investment banking can underwrite debt issues, both on their own behalf and for corporate and public sector clients, as well as supply equity, manage funds and offer advice to corporate mergers and acquisitions. These banks act as financial intermediaries, raising capital (equity and debt) by trading foreign exchange, commodities and equity securities on the primary market.

3) Insurance

The insurance sector can generally be divided into two categories:
Life Insurance; and General (Non-Life) Insurance.
"Green" insurance falls under the latter and typically encompasses two product areas:
1) those which allow an insurance premium differentiation on the basis of environmentally relevant characteristics; and
2) Insurance products specifically tailored for clean technologies and emissions reducing activities.

4) Asset Management

Asset Management has become one of the fastest growing segments in the financial industry and represents a core business unit of current banks. This space focuses on providing financial advice to clients on estate planning, mutual funds, managed asset programs, taxes, trust services, international financial planning, global private banking and full-service and discount brokerages.

Green Retail Banking
Ask something more from your local Bank

 Home Mortgage	• 'green' mortgage initiative. %1 reduction on interest • for loans that meet environmental criteria. • Free home energy rating and offsets carbon emissions for every year of loan • Green Power Oriented Mortgage • 10% premium refund on its mortgage loan insurance	
 Commercial Building Loan	• Green Loans for new condos • Developers do not have to pay an initial premium for "green" commercial buildings • Provides 1/8 of 1% discount on loans to green leadership projects	
 Auto Loan	Clean Air Auto Loan with preferential rates for hybrids & Electrical	
 Credit Card	Climate Credit Card. Bank will donate to WWF GreenCard Visa is the world's first credit card to offer an emissions offset program	
 Deposit	Fully-insured deposits earmarked for lending to local energy-efficient companies aiming to reduce waste/pollution, or conserve natural resources	

Green Corporate and Investment Banking

Ask something more from your local Bank

 Project Finance	Specialized service divisions are dedicated to long-term financing of clean energy projects. Some banks also specialize in one (or several) renewable technology type and/or place a premium on working with states where regulatory framework and government policy encourages the early adoption of clean technologies.
 Securitization	A risk sharing arrangement for environmental projects. Financial institution represents a guarantor (or structuring investor) at the mezzanine level of risk, allowing client to transfer risk to bank. Eco-Securitization scheme will test the feasibility of financing "natural infrastructure" by linking sustainable management of resources with the funding capacity and requirements of asset-backed securitization.
 Carbon Finance	Banks provide equity, loans and/or upfront or upon delivery payments to acquire carbon credits from projects. Most acquire carbon credits in order to serve their corporate clients' compliance needs, supply a tradable product to the banks' trading desks, or develop lending products backed by emission allowances and carbon credits.
 Indices	Series of environmental private investor eco-market products includes a biofuels commodity basket, total returns solar energy index, clean renewable energy index and total returns water index (e.g., enables interested parties to invest in water as a commodity).
 Deposit	Fully-insured deposits earmarked for lending to local energy-efficient companies aiming to reduce waste/pollution, or conserve natural resources

Insurance

Ask something more from your local Insurance Co.

 Auto Insurance	Pay As You Drive™ Insurance. Mileage-based Insurance. 10% discount for hybrid and fuel efficient vehicles. Bank can also choose to offset vehicle's annual emissions (e.g. 20 % emissions offset through Climate Care Recycling Insurance. Customer pays less for car insurance, by up to 20%, if recycled parts are used when vehicle is damaged and requires service.
 Building and Home Insurance	Green Building Replacement and Upgrade Coverage. Product covers unique type of "green" risks related to the sustainable building industry. "Climate Neutral" Home Insurance Policy. First home insurance product to carry out GHG offsetting based on customer usage.
 Business Insurance	Environmental Damage Insurance.

Asset Management

Ask something more from your local Financial Organization

 Fisacl Green Fund	By purchasing shares or investing in Green Funds, customers receive an income tax discount, and thus accept a lower interest rate on investment. Banks can offer loans at lower cost to finance environmental projects related to eligible categories.
 Fund	Eco Performance is the world's largest "green" fund. %80 of assets are channeled towards eco and social leaders, with %20 going to "eco-innovators". Equity Fund - Future Energy, focuses on clean energy sector investments in clean four energy-related business segments.
 Cat Bond Fund	Cat Bond Fund. World's first public fund for catastrophe bonds, a portion of which is aimed at climate-related natural disasters (or climate adaptation). Vehicle designed to hedge climate risks typically difficult to cover in the traditional insurance market.

Conclusion
Many "green" financial products and services, reviewed above, either remain in the nascent stage of development/implementation or data related to their success/failure has not yet been generated or reported.
Due to this lack of experience and data, any rigorous measurement or ranking of these designs would be overly speculative and risk misrepresenting some designs over others. Looking ahead, however, as more quantitative and qualitative track records emerge for these products,

The following questions should be considered when gauging product performance or promise:
- Does it achieve high levels of financial performance?
- Does it attract a particularly large number of customers?
- Does it last over time, and is re-launched year by year?
- Does it raise the environmental awareness among all stakeholders, including clients and employees?
- Does it receive positive attention from the media and environmental NGOs?
- Does it prompt the introduction of other environmental products and services?
- Does it improve brand recognition and corporate image among stakeholders?

INTERNATIONAL MANUAL OF CLIMATE CHANGE CONTROL (IMCCC) • 233

As environmental understanding and awareness grows in North America, so too will the demand for products and services aimed at facilitating the advancement of environmentally sustainable lives, livelihoods and communities. At the same time, this demand will also expose new business opportunities, while leading to an increased diversification of products and services found in multiple sectors. Consequently, organizations that have the foresight and capacity to tap into this desire by consumers to affect positive environmental change will likely experience widespread benefits; from improved corporate image to increased growth and competitiveness in the marketplace. Given their intermediary role in the economy and farreaching customer base, financial institutions will be well-positioned to reap financial and non-financial rewards, while furthering their contribution toward sustainable development.

Are Crypto Currencies Inherently Bad For The Environment?
All cryptocurrencies have an energy and environmental problem. But if done right, it might be possible to channel all that energy into something good for the planet.

Crypto's environmental troubles

A fierce debate around the environmental impacts of cryptocurrencies, like bitcoin, is growing. Bitcoin does consume a lot of energy. That energy use is growing and annually consumes as much energy as whole nations, such as Finland, Malaysia, or Sweden. While bitcoin is not the only industry to consume as much energy as entire countries, e.g. concrete consumes more energy than India, the energy both sectors consume comes with associated pollution, including carbon emissions.

Even transactions with bitcoin use a lot of energy, with the average transaction consuming over 1,700 kWh of electricity, which is almost twice the monthly amount used by the average U.S. home. However, there was ways to transact in bitcoin using much less energy.

Exacerbating this problem, some bitcoin mining operations have teamed up with struggling fossil fuel power plants, keeping some power plants online that would otherwise have retired, increasing overall carbon emissions. Some utilities have even gotten into the bitcoin game directly.

Large bitcoin mining operations are also moving locations as China, the country previously with the largest bitcoin mining industry, recently banned both cryptocurrency mining and transactions. This change has bitcoin mining operations moving to places like Texas and potentially Alberta, Canada.

All else equal, bitcoin operations that co-locate and utilize fossil fuels that would have otherwise stayed in the ground will increase emissions.

Some are considering using stranded natural gas that would otherwise have been flared, which, absent any methane venting and flaring regulations, would make the use of the natural gas for bitcoin, at-best, carbon neutral. However, it is a stretch, and making the natural gas more valuable at the wellhead could further dissuade pipeline development that would have moved the gas to market.

However, co-locating bitcoin mining operations with zero-carbon resources, such as nuclear, hydro, wind and solar, could help reduce the carbon emissions associated with the mining itself. Co-location could also give a financial boost to power plants that might be able to sell their electricity at a higher price to miners instead of to the grid when demand and prices are low. This type of hybrid power plant/mine might even make uneconomical projects economical.

Going further, it is also possible that the cryptocurrency mines themselves could offer benefits directly to the grid, and, if operated intelligently, even result in lower overall carbon emissions.

A positive grid impact?

The study simulated the evolution of the Electric Reliability Council of Texas (ERCOT), the grid that serves most of Texas, out to 2030 under multiple scenarios:
1) a base case with no datacenter/bitcoin mining expansion,
2) a case with 5 GW of inflexible (always on) datacenter/bitcoin mines by 2030,
3) a scenario with 5 GW of mildly flexible datacenters deployed by 2030 and
4) a scenario with 5 GW of very flexible datacenters deployed by 2030.

The non-flexible scenario added a significant baseload to the ERCOT system. This growth resulted in the deployment of more power plant capacity that the base case, including more wind, natural gas, and solar.
This increased energy use also resulted in an additional 7.9 million metric tons of carbon emissions over the base case by 2030.

However, the flexible scenarios were more interesting. Both flexible scenarios actually see more wind and less natural gas deployed than both the base case and the inflexible scenarios.

This change is because the datacenters/mines were programed to reduce their energy consumption by certain percentages when electricity prices hit certain tiers. In total, the third scenario saw the datacenters/mines curtailing their load about 14% of the year.

The flexibility of the datacenters/mines in the latter two scenarios allowed the model to deploy different levels of technologies than the base or inflexible case. The model actually built more renewables because it could utilize the flexibility of the datacenters/mines to compensate for fluctuations in renewable output. This flexibility also resulted in lower carbon emissions compared to the base case.

For additional load to result in lower total carbon emissions, the additional energy consumption must be offset by more zero-carbon energy. In the flexible datacenter/mine cases, the amount of energy generated from wind and solar was more than in the base case and the amount generated from natural gas was lower.

In general, the flexibility of the datacenters/mines moves their load to more value energy over power, which better aligns with renewables. This is because renewables are great at providing large amounts of energy, but have less ability to always provide capacity, or constant power.

In concept, flexible datacenters/mines are similar to the electrification of transportation or heating with the ability to control then times when the chargers and heaters operate. However, it is likely that datacenters/mines could offer large levels of flexible load concentrated in a smaller number of locations, which could make their administration easier.

Grid decarbonization studies often assume high levels of flexible demand, and often much of this flexibility comes from diffuse sources, such as smart thermostats and EV charging. While this analysis did not seek to satisfy any carbon policy, it does illustrate the potential carbon benefits of high levels of flexible demand coupled with an electricity market that is able to incorporate it.

Mining and transacting cryptocurrencies, such as bitcoin, do present energy and emissions challenges, but new research shows that there are possible pathways to mitigate some of these issues if cryptocurrency miners are willing to operate in a way to compliment the deployment of more low-carbon energy.

The author of this Book does not currently own or mine any cryptocurrencies.

Step Eleven

Tourism

STEP ELEVEN

All about 'Eco-Tourism'

The term Ecotourism emerged in the late 1980s as a direct result of theworld's acknowledgment and reaction to sustainable practices and global ecological practices. In these instances, the natural-based element of holiday activities together with the increased awareness to minimise the 'antagonistic' impacts of tourism on the environment (which is the boundless consumption of environmental resources) contributed to the demand for ecotourism holidays. This demand was also boosted by concrete evidence that consumers had shifted away from mass tourism towards experiences that were more individualistic and enriching. In addition, these experiences were claimed to be associated with a general search for the natural component during holidays

Definitions of Ecotourism

Ziffer, 1989	'Ecotourism is a form of tourism inspired primarily by the natural history of an area, including its indigenous cultures. The ecotourist visits relatively undeveloped areas in the spirit of appreciation, participation and sensitivity. The ecotourist practices a non-consumptive use of wildlife and natural resources and contributes to the visited area through labor or financial means aimed at directly benefiting the conservation of the site and the economic well-being of the local residents...'
Boo, 1991	'Ecotourism is a nature tourism that contributes to conservation, through generating funds for protected areas, creating employment opportunities for local communities, and offering environmental education.'
Forestry Tasmania, 1994	'Nature-based tourism that is focused on provision of learning opportunities while roviding local and regional benefits, while demonstrating environmental, social, cultural, and economic sustainability'
Richardson, 1993	'Ecologically sustainable tourism in natural areas that interprets local environment and cultures, furthers the tourists' understanding of them, fosters conservation and adds to the well-being of the local people.'
Australia Department of Tourism, 1994	'Nature-based tourism that involves education and interpretation of the natural environment and is managed to be ecologically sustainable. This definition recognizes that natural environment includes cultural components, and that ecologically sustainable involves an appropriate return to the local community and long-term conservation of the resource.'
Figgis, 1993	'Travel to remote or natural areas which aims to enhance understanding and appreciation of natural environment and cultural heritage, avoiding damage or deterioration of the "environment and the experience for others".'
Tickell, 1994	'Travel to enjoy the world's amazing diversity of natural life and human culture without causing damage to either.'

Definitions of Ecotourism (Cont.)

Boyd & Butler, 1993	'A responsible nature travel experience, that contributes to the conservation of the ecosystem while respecting the integrity of host communities and, where possible, ensuring that activities are complementary, or at least compatible, with existing resource- based uses present at the ecosystem.'
Boyd & Butler, 1996	'Ecotourism is a form of tourism which fosters environmental principles, with an emphasis on visiting and observing natural areas'
Goodwin, 1996	'Low impact nature tourism which contributes to the maintenance of species and habitats either directly through a contribution to conservation and/or indirectly by providing revenue to the local community sufficient for local people, and therefore protect, their wildlife heritage area as a source of income.'
Lindberg & McKercher, 1997	'Ecotourism is tourism and recreation that is both nature-based and sustainable.'

Environmental Impacts of Ecotourism

The most proclaimed positive issue is ecotourism's contribution to sustainable resource management through conservation of the natural resources on a direct or indirect basis (Commonwealth of Australia, 1993, 1995; Cater, 1993, 1994; Dearden, 1995)

Environmental impacts	
Direct benefits	Direct costs
• Provides incentive to protect environment, both formally (protected areas) and informally	• Danger that environmental carrying capacities will be unintentionally exceeded, due to:
• Provides incentive for restoration and conversion of modified habitats	• Rapid growth rates Difficulties in identifying, measuring and monitoring impacts over a long period
• Ecotourists actively assisting in habitat enhancement (donations, policing, maintenance, etc.)	• Idea that all tourism induces stress

Environmental impacts (Cont.)	
Indirect benefits	Indirect costs
• Exposure to ecotourism fosters broader commitment to environmental well-being	• Fragile areas may be exposed to less benign forms of tourism (pioneer function)
• Space protected because of ecotourism provide various environmental benefits	• May foster tendencies to put financial value on nature, depending upon attractiveness

Economic Impacts of Ecotourism

The direct and indirect benefits which are derived from biodiversity conservation, represent the fundamental goal of ecotourism, by attracting visitors to the natural settings and using the revenues to fund conservation and fuel economic development (Commonwealth of Australia, 1995: 12; Cater, 1993, 1994)

Economic impacts	
Direct benefits	Direct costs
• Revenues obtained directly from ecotourists • Creation of direct employment opportunities • Strong potential for linkages with other sectors of the local economy • Stimulation of peripheral rural economies	• Start-up expenses (acquisition of land, establishment of protected areas, superstructure, infrastructure) • Ongoing expenses maintenance of infrastructure, promotion, wages)

Economic impacts (Cont.)	
Indirect benefits	Indirect costs
• Indirect revenues from ecotourists (high multiplier effect) • Tendency of ecotourists to patronise cultural and heritage attractions as 'add-ons' • Economic benefits from sustainable use of protected areas and inherent existence	• Revenue uncertainties to in situ nature if consumption • Revenue leakages due to imports, expatriate or non-local participation, etc. • Opportunity costs • Damage to crops by wildlife

Sociocultural Impacts of Ecotourism

The sustainable component of ecotourism often attests certain direct and indirect sociocultural benefits and costs at the sites and/or at the destination level . Generally speaking, it was proposed that the assessment of the cultural impacts of ecotourism could be based on four criteria , commodification element; culture affecting social change; cultural knowledge; and cultural patrimony elements.

Sociocultural impacts	
Direct benefits	Direct costs
• Ecotourism accessible to a broad spectrum of the population • Aesthetic/spiritual element of experiences • Foster environmental wareness among ecotourists and local population	• Intrusions upon local and possibly isolated cultures • Imposition of elite alien value system • Displacement of local cultures by parks • Erosion of local control (foreign experts, in-migration of job seekers).

Sociocultural impacts (Cont.)	
Indirect benefits	Indirect costs
• Option and existence benefits	• Potential resentment and antagonism of locals • Tourist opposition to aspects of local culture (e.g. hunting, slash-burn agriculture).

Principles of Ecotourism

(1) travel to natural destinations.

(2) minimizes impact. This includes minimizing the impact of development and tourist activity by choosing appropriate building materials, renewable energy sources, visitor management strategies, monitoring techniques and conservation plans.

(3) builds environmental awareness. This includes educational and interpretational material for visitors, educational training for guides and educating the greater public and surrounding community.

(4) provides direct financial benefit for conservation.

(5) provides financial benefits and empowerment for local people. This includes employment of local people, using an all-inclusive stakeholder approach to planning, management and policy development and fostering of partnerships.

(6) respects local culture.

(7) supports human rights.

Step Twelve: Knowledge Test & Certification

ONLINE EDUCATION

START YOUR ONLINE COURSE NOW AFTER YOU HAVE COMPLETED READING THIS BOOK, DO YOUR QUIZZES AND RECEIVE YOUR
INTERNATIONAL TRAINING CERTIFICATE OF :
CLIMATE CHANGE CONTROL

JOIN US AT:

WWW.TOPTENAWARD.ORG

Bibliography:

Achama, F. (1995) Defining ecotourism. In L. Haysith and J. Harvey (eds) Nature Conservation and Ecotourism in Central America (pp. 23–32). Florida: Wildlife Conservation Society.

Agardy, M.T. (1993)Accommodating ecotourism in multiple use planning of coastal and marine protected areas. Ocean & Coastal Management 20 (3), 219–239.

Asadi, J. (2021-22) International Environmental Labelling (Vol.1-11).

Australia Department of Tourism (1994)National Ecotourism Strategy. Canberra: Australia Government Publishing Service.

Ayala, H. (1995) From quality product to Eco-product: Will Fiji set a precedent? Tourism Management 16 (1), 39–47.

Barnes, J.L. (1996)Economic characteristics of the demand for wildlife-viewing tourism in Botswana. Development Southern Africa 13 (3), 377–397.

Blamey, R.K. (1995a) The Nature of Ecotourism. Canberra: Bureau of Tourism Research.

Blamey, R.K. (1995b) The elusive market profile: Operationalization ecotourism. Paper presented at the Geography of Sustainable Tourism Conference, University of Canberra, ACT, Australia, September.

Blamey, R.K. (1997) Ecotourism: The search for an operational definition. Journal of Sustainable Tourism 5 (2), 109–130.

Boo, E. (1990) Ecotourism: The Potential and Pitfalls (Vols 1& 2). Washington, DC: World Wide Fund for Nature.

Boo, E. (1991a) Ecotourism: A tool for conservation and development. In J.A. Kusler (compiler) Ecotourism and Resource Conservation: A Collection of Papers (Vol. 1) (pp. 54–60). Madison: Omnipress.

Boo, E. (1991b) Planning for ecotourism. Parks 2 (3), 4–8.

Boo, E. (1992) The Ecotourism Boom: Planning for Development and Management. WHN technical paper series, Paper 2. Washington, DC: WWF.

Boo, E. (1993) Ecotourism planning for protected areas. In K. Lindberg and D.E. Hawkins (eds) Ecotourism: Guide for Planners and Managers (pp. 15–31). North Bennington: The Ecotourism Society.

Bottrill C.G. and Pearce, D.G. (1995) Ecotourism: Towards a key elements to operationalising the concept. Journal of Sustainable Tourism 3 (1), 45–54.

Amberg, N.; Magda, R. Environmental Pollution and Sustainability or the Impact of the Environmentally Conscious Measures of International Cosmetic Companies on Purchasing Organic Cosmetics. Visegrad J. Bioecon. Sustain. Dev. 2018, 1, 23.

Asadi, J., "International Environmental Labelling, Economic Consequences, Export Magazine, July 2001

Asadi, J. 2008. Mobile Phone as management systems tools, ISO Magazine, Vol.8, No.1

Asadi, J., Eco-Labelling Standards, National Standard Magazine, Sep. 2004.

Barbieux, D.; Padula, A.D. Paths and Challenges of New Technologies: The Case of Nanotechnology-Based Cosmetics Development in Brazil. Adm. Sci. 2018, 8, 16.

Advanced Engineering and Applied Sciences: An International Journal 2014; 4(3): 26-28

Berolzheimer, C. (2006). Pencils: An Environmental Profile.

Chemical Week, 1999. Europe's Beef Ban Tests Precautionary Principle. (August 11).

Chaudri, S.K.; Jain, N.K. History of Cosmetics. Asian J. Pharm. 2009, 7–9, 164–167.

CHOI, J.P. Brand Extension as Informational Leverage. Review of Eco- nomic Studies, Vol. 65 (1998), pp. 655-669.

Conway, G. 2000. Genetically modified crops: risks and promise.

Corrado, M., (1989), The Greening Consumer in Britain, MORI, London

Corrado, M., (1997), Green Behaviour – Sustainable Trends, Sustainable Lives?, MORI, london, accessed via countries. Manila, Asian Development Bank 33p.

Davies, Clive. Chief, Design for the Environment Program, Environmental Protection Agency. Interview. March 24, 2009.

Federal Trade Commission, "Sorting Out Green Advertising Claims." http://www.ftc.gov/bcp/edu/pubs/consumer/general/gen02.shtm (March 26, 2009, March 27, 2009)

Ooyen, Carla. Research Manager with Nutrition Business Journal. Personal correspondence. March 19, 2009.

Tekin, Jenn. Marketing Manager with Packaged Facts & SBI. Personal correspondence. March 17, 2009.

University of California - Berkeley. http://berkeley.edu/news/media/releases/2006/05/22_householdchemicals.shtml (March 26, 2009)

U.S. Department of Health and Human Services, Household Products Database.http://householdproducts.nlm.nih.gov/cgi-bin/household/prodtree?prodcat=Inside+the+Home (March 17,

Women's Voices of the Earth, "Household Cleaning Products and Effects on Human Health."http://www.womenandenvironment.org/campaignsandprograms/SafeCleaning/safecleaninghealth (March 17, 2009)

EMONS, W. Credence Goods Monopolists. International Journal of In- dustrial Organization, Vol. 19 (2001), pp. 375-389.

European Union official website: https://ec.europa.eu/info/about-european-commission/contact_en

Feenstra, R.C. "Exact Hedonic Price Indexes," Review of Economics and Statistics 77 (1995): 634-653.

Feenstra, R.C., and J.A. Levinsohn. "Estimating Markups and Market Conduct with Multidimensional Product Attributes," Review of Economic Studies (62 (1995): 19-52.

ForestEthics. (n.d.). Back to School Report Card.

Forest Stewardship Council: "Principles and criteria for forest stewardship" Document 1.2: <http://www.fscoax.org>

Forsyth, K. 1999. Will consumers pay more for certified wood products? Journal of Forestry 97 (2) : 18-22.

Forest Choice #2 (2014, January 1). Forest Choice #2 Graphite Pencils (12 Pack).

Francois, C., Harris, B. (2014, November 2). How are Mechanical Pencils Made?.

Freeman, A. M III. The Measurement of Environmental and Resource Values. Theory and Methods. Washington D.C.: Resource for the Future, 1993.

Friends of the Earth, 1993. Timber certification and eco-labeling. London, FOE:

Geetha Margret Soundri, "Ecofriendly Antimicrobial Finishing of Textiles Using Natural Extract", Journal of International Academic Research For Multidisciplinary, ISSN: 2320 – 5083, 2014, Vol 2.

Graves, P., J.C. Murdoch, M.A. Thayer, and D. Waldman. "The Robustness of Hedonic Price Estimation: Urban Air Quality," Land Economics 64(1988): 220-233.

Halvorsen, R. and R. Palmquist. "The Interpretation of Dummy Variables in Semilogarithmic Equations." American Economic Review 70:474-75 (1980).

Henderson D. (2008). Opportunity Cost." The Concise Encyclopedia of Economics.

How It's Made. (2009, Nov 17). How It's Made Graphite Pencil Leads [video file].

Imhoff, Dan. "Growing Pains: Organic Cotton Tests the Fibre of Growers and Manufacturers Alike," reprinted on Simple Life's web page (simplelife.com), but first printed by Farmer to Farmer, December 1995.

Incomplete Consumer Information in Laboratory Markets. Journal of Environmental labeling.

ISO 14020, ISO 14021,ISO 14024,ISO 14025, International Organization for Standardization.

Kennedy, P.E. "Estimation with Correctly Interpreted Dummy Variables in Semilogarithmic Equations," American Economic Review 71: 801 (1981).

Kirchho®, S., (2000), Green Business and Blue Angels.

Kraus, Jeff. Lab Technician at the North Carolina School of Textiles.

Labeling Issues, Policies and Practices Worldwide.

Lamport, L. 1998. The cast of (timber) certifiers: who are they? International J. Ecoforestry 11(4): 118-122.

Large Scale impoverishment of Amazonian forests by logging and fire. 1999.

Lathrop, K.W. and Centner, T.J. 1998. Eco-labeling and ISO 14000: An analysis of US regulatory systems and issues concerning adoption of type II standards. Environmental

Lee, J. et al. 1996. Trade related environmental measures; sizing and comparing impacts.

Lehtonen, Markku. 1997. Criteria in Environmental Labeling: A comparative Analysis on Environmental Criteria in Selected Labeling Schemes. Geneva, UNEP. 148p.

LIEBI, T. Trusting Labels: A Matter of Numbers? Working Paper Uni versity of Bern, No. 0201 (2002).

OECD. "Ec-labelling: Actual Effects of Selected Programmes," OCDE/GD (97) 105, 1997, Paris. (Available on line at http://www.oecd.org/env/eco/books.htm#trademono)

OECD. 1997a. Case study on eco-labeling schemes. Paris, OECD (30 Dec):

OECD. 1997b. Eco-labeling: Actual Effects of Selected Programs.

Osborne, L. "Market Structure, Hedonic Models, and the Valuation of Environmental Amenities." Unpublished Ph.D. dissertation. North Carolina State University, 1995.

Osborne, L., and V. K. Smith. "Environmental Amenities, Product Differentiation, and market Power," Mimeo, 1997.

Ozanne, L.K. and Vlosky, R.P. 1996. Wood products environmental certification: the United States perspective". Forestry Chronicle 72 (2) : 157-165.

Palmquist, R. B., F. M. Roka, and T.Vukina. "Hog Operations, Environmental Effects, and Residential Property Values," Land Economics 73(1), (1997): 114-24.

Palmquist, R.B. "Hedonic Methods," in J.B Braden and C.D. Kolstad, eds. Measuring the Demand for Environmental Improvement. Amsterdam, NL: Elsevier, 1991.

Paper Mate. (2014). Paper Mate Recycled.

Pento, T. 1997. Implementation of Public Green Procurement Programs (22-31) in Greener Purchasing: Opportunities and Innovations. Sheffield, Greenleaf Publ. 325 p.

Perloff, J. "Industrial Organization Lecture Notes," Mimeo. University of California at Berkeley (1985).

Plant, C. and Plant, J. 1991. Green business: hope or hoax? Philadelphia, New Society Publishers 136 p.

Pencil Making Today (2014, January 1). Pencil Making Today: How to Make a Pencil in 10 Steps.

Polak, J. and Bergholm, K. 1997. Eco-labeling and trade: a cooperative approach (Jan.): Policy in a Green Market. Environmental and Resource Economics 22, 419-

Poore, M.E.D. et al. 1989. No timber without trees. London, Earthscan. 352p.

Raff, D. M.G., and M. Trajtenberg. "Quality-Adjusted Prices for the American Automobile Industry: 1906-1940." NBER Working Paper Series, Working Paper No. 5035, February 1995.

Roberts, J. T. 1998. Emerging global environment standards: prospects and perils. Journal of Developing Societies 14 (1): 144-163.

Rosen, S., "Hedonic Prices and Implicit Markets: Product Differentiation in Pure Competition." Journal of Political Economy. 82: 34-55 (1974).

Ross, B. 1997. Eco-friendly procurement training course for UN HCR. : 126 p.

Sayre, D. 1996. Inside ISO 14000: The competitive advantage of environmental management. Delray Beach FL., St. Lucie Press. 232p.

Suzuki, D. (2014, January 1). PEG Compounds and their contaminants

SHAPIRO, C. Premiums for High Quality Products as Returns to Reputa- tion. Quarterly Journal of Economics, Vol. 98, No. 4 (1983), pp. 659-680.

Stillwell, M. and van Dyke, B. 1999. An activists handbook on genetically modified organisms and the WTO. Washington DC., The Consumer's Choice Council: 20 p.

Semenzato, A.; Costantini, A.; Meloni, M.; Maramaldi, G.; Meneghin, M.; Baratto, G. Formulating O/W Emulsions with Plant-Based Actives: A Stability Challenge for an Eective Product. Cosmetics 2018, 5, 59.

Sources of Plastics (2014, January 1). Sources of Plastics.

Singh, S. (2008, March 6). Paraffin wax.

Saint Jean Carbon. (n.d.). Sri Lankan Graphite.

U.S. Environmental Protection Agency. National Water Quality Fact Inventory: 1990 Report to Congress. EPA 503-9-92-006, Apr. 1992.

UK Eco-labelling Board website, accessed via http://www.ecosite.co.uk/Ecolabel-UK/

US Environmental Protection Agency (EPA742-R-99-001): 40 p. <www.epa.gov/opptintr/epp>

US EPA, 1993. Determinants of effectiveness for environmental certification and labeling programs. Washington, D.C., US Environmental Protect

US EPA, 1993. Status report on the use of environmental labels worldwide. Washington, D.C., US Environmental Protection Agency (742-R-93-001 September).

US EPA, 1993. The use of life-cycle assessment in environmental labeling. Washington, D.C., US Environmental Protection Agency (742-R-93-003 September).

US EPA, 1998. Environmental labeling: issues, policies, and practices worldwide. Washington DC., Environmental Protection Agency, Pollution Prevention Division Prepared by Abt

USG, 1998. Greening the government through waste prevention, recycling, and federal acquisition. Washington, D.C., Executive Order 13101 (September).

Kijjoa, A.; Sawangwong, P. Drugs and Cosmetics from the Sea. Mar. Drugs 2004, 2, 73–82. [CrossRef]

Wang, J.; Pan, L.; Wu, S.; Lu, L.; Xu, Y.; Zhu, Y.; Guo, M.; Zhuang, S. Recent Advances on Endocrine Disrupting Eects of UV Filters. Int. J. Environ. Res. Public Health 2016, 13, 782.

Bilal, A.I.; Tilahun, Z.; Shimels, T.; Gelan, Y.B.; Osman, E.D. Cosmetics Utilization Practice in Jigjiga Town, Eastern Ethiopia: A Community Based Cross-Sectional Study. Cosmetics 2016, 3, 40.

Ting, C.T.; Hsieh, C.M.; Chang, H.-P.; Chen, H.-S. Environmental Consciousness and Green Customer Behavior: The Moderating Roles of Incentive Mechanisms. Sustainability 2019, 11, 819.

Chen, K.; Deng, T. Research on the Green Purchase Intentions from the Perspective of Product Knowledge. Sustainability 2016, 8, 943.

Wang, H.; Ma, B.; Bai, R. How Does Green Product Knowledge Eectively Promote Green Purchase Intention? Sustainability 2019, 11, 1193.

Nguyen, T.T.H.; Yang, Z.; Nguyen, N.; Johnson, L.W.; Cao, T.K. Greenwash and Green Purchase Intention: The Mediating Role of Green Skepticism. Sustainability 2019, 11, 2653.

Cinelli, P.; Coltelli, M.B.; Signori, F.; Morganti, P.; Lazzeri, A. Cosmetic Packaging to Save the Environment: Future Perspectives. Cosmetics 2019, 6, 26.

Eixarch, H.; Wyness, L.; Siband, M. The Regulation of Personalized Cosmetics in the EU. Cosmetics 2019, 6, 29.

CANADA BRONZE BEAVER BADGE

Participate in our Online Classes to earn these exclusive digital badges!

Design & Development by:

Tara Asadi

Copyright © 2022 by Top Ten Award International Network.

CANADA SILVER BEAVER BADGE

Participate in our Online Classes to earn these exclusive digital badges!

Design & Development by:

Tara Asadi

Copyright © 2022 by Top Ten Award International Network.

CANADA GOLD BEAVER BADGE

Participate in our Online Classes to earn these exclusive digital badges!

Design & Development by:

Tara Asadi

Copyright © 2022 by Top Ten Award International Network.